Alternative Swine Management Systems

Alternative Swine Management Systems

Ioan Hutu, PhD

Assistant Professor
Department of Animal Production and Veterinary Public Health
Banat University of Agricultural Science and Veterinary Medicine
The King Michael I of Romania, Timisoara, Romania

Gary Onan, PhD

Professor
Department of Animal and Food Science
University of Wisconsin, River Falls
Wisconsin, United States

ELSEVIER

ACADEMIC PRESS
An imprint of Elsevier

Academic Press is an imprint of Elsevier
125 London Wall, London EC2Y 5AS, United Kingdom
525 B Street, Suite 1650, San Diego, CA 92101, United States
50 Hampshire Street, 5th Floor, Cambridge, MA 02139, United States
The Boulevard, Langford Lane, Kidlington, Oxford OX5 1GB, United Kingdom

ALTERNATIVE SWINE MANAGEMENT SYSTEMS

Notices
Practitioners and researchers must always rely on their own experience and knowledge in
evaluating and using any information, methods, compounds or experiments described
herein. Because of rapid advances in the medical sciences, in particular, independent
verification of diagnoses and drug dosages should be made. To the fullest extent of the
law, no responsibility is assumed by Elsevier, authors, editors or contributors for any injury
and/or damage to persons or property as a matter of products liability, negligence or
otherwise, or from any use or operation of any methods, products, instructions, or ideas
contained in the material herein.

ISBN: 978-0-12-818967-2

Publisher: Charlotte Cockle
Acquisition Editor: Anna Valutkevich
Editorial Project Manager: Theresa Yannetty
Production Project Manager: Kiruthika Govindaraju
Cover Designer: Alan Studholme

Contents

Foreword

The Romanian version of the book offered some alternative production strategies that may fit well with Romania's swine production traditions. These alternative technologies result in productivity on a par with current standard systems and are more cost-effective. In addition, they offer an approach that considers the concerns of consumers who are increasingly interested not just in food but also the way in which it was produced. These alternative systems create a very animal welfare friendly environment, and producers can maintain a transparency within their production practices that modern proactive consumers will find desirable. Therefore these alternative production systems offer a tremendous opportunity for Romanian pork producers to produce superior products in a very cost-effective manner. It remains to be seen if producers will adopt these systems. We realize that there are numerous barriers to adoption including an unwillingness to change and a lack of education among many producers. It is our sincere hope that producers will adopt these technologies and adapt them to their benefit. Time will tell.

The initial impetus for this work came from a symposium on the use of tarp-covered hoop structures for swine production held at Iowa State University in Ames, IA, United States, in September 2004. Much of what appears within this book are materials originally presented at that symposium. We would like to particularly acknowledge the contributions of Professor Mark Honeyman to this work as well as all the many other scientists whose materials such as tables, charts, and pictures we have used. This book is also in a large part the result of cooperative endeavors between the University of Wisconsin-River Falls, United States, and Banat's University of Agricultural Science and Veterinary Medicine from Timişoara, Romania. These have included a presentation by Dr. Onan at the annual Veterinary Conference in Timisoara in 2005 as well as a number of articles by Dr. Hutu using resources from River Falls in a variety of venues such as *Ferma* magazine, *Agroinfo* forum (www.agroinfo.ro) and through the *Unit of Extension Service of Banat's University of Agricultural Science and Veterinary Medicine at Timisoara* (www.unitate-extensie.org.ro).

The majority of the information in this book is the result of on-site research performed in Romania and the United States. Understandably most of the research results are from the United States, and our focus has been to use data from areas within the United States whose climate and soils are similar to those in Romania, namely, the states of Iowa, Minnesota, and Wisconsin. We feel the information given is very valid for Romanian swine production and hope that it will stimulate the industry in Romania and ultimately promote joint public—private research within Romania that will optimize these technologies for that country's specific needs.

The English version of the book, *Alternative Swine Management Systems,* was edited after the construction of the Swine Experimental Unit at *Horia Cernescu* Research Complex, a part of Banat University of Agricultural Science and Veterinary Medicine in Timisoara, Romania. It was a portion of an infrastructure development project—*Development of research, education and services infrastructure in the fields of veterinary medicine and innovative technologies for Romania West Region*—financed by the European Union in order to promote research and extension activities.

Revisions for the second edition also began shortly after Dr. Gary Onan completed a 5-month sabbatic at Banat University of Agricultural Science and Veterinary Medicine. While there, he taught courses within the Animal Production Department of the Veterinary School and worked with their Extension Unit. He and Dr. Hutu conducted a grow-finish demonstration trial in the Swine Experimental Unit during that time to promote production approach among Romanian farmers and agricultural students.

A humorous anecdote regarding that trial and the Banat Agralim Fair occurred that spring. A "serious problem" occurred, whereby the food court for the Fair would be immediately adjacent to the swine unit. The Fair administrators were concerned about bad publicity arising from the smell of the swine so near the food vendors. The Fair officials wanted the pigs removed, even though the trial was ongoing. After dozens of phone calls and numerous site visits by Fair officials, it was decided the pigs could stay, as long as the researchers assume responsibility for any bad publicity. As it turned out, the Swine Experimental Unit was the most visited site at the exhibition and the university was deprived of the opportunity for a "beautiful" public relations disaster. This story demonstrates that social acceptance of this type of system is a point that should not be neglected, even among intellectuals such as university faculty and students.

Gary Onan and Ioan Hutu during Banat Agralim Fair.

Swine Experimental Units, 2016, Banat's University of Agricultural Science and Veterinary Medicine from Timisoara, Romania.

Future goals for the swine research group are to use the Swine Experimental Unit for studies focused on use of bacteriophages, probiotics, and natural extracts as adjuncts for maintenance of swine health in this type of system. From the perspective of the Animal Production Extension staff, the use of this structure will help increase interest and demonstrate applicability of this economically efficient and environmentally sustainable system to swine producers.

It is our hope that the detailed information and pictures of the construction of the Swine Experimental Unit, and its subsequent use, which have been added to the English edition, will encourage the Romanian farmers, as well as those in other world regions undergoing animal agricultural development, to adapt this low capital investment, low energy input, environmentally friendly system. Realizing that much work yet remains to develop swine production technologies that are particularly suited to regions in agricultural development, we welcome any discussion or feedback that may stimulate further investigation and discourse.

Authors
Timisoara and River Falls
February 2019

Acknowledgments

We are grateful to the following persons for permission to reproduce copyright material:

Professor *Mark Honeyman* from Iowa State University and Leopold Center for Sustainable Agriculture who gave permission to use the data, pictures, and materials presented at the National Conference on Hoop Barns and Bedded Systems for Livestock Production, held on September 14–15, 2004. The conference materials provided data, designs, and pictures presented by speakers and moderators including *Hugh Payne* (Animal Industries, Department of Agriculture Western Australia), *Artur Luza* (Agro-Soyuz, Ukraine), *John Maltman* (Manitoba Agriculture, Food and Rural Initiatives, Canada), *Maynard Hogberg* (Animal Science Department, ISU*), Peter Lammers* (Animal Science Department ISU).

Professor Emeritus *Bo Algers*, Swedish University of Agricultural Sciences, Skara, Sweden, for pictures regarding the Swedish group farrowing system.

Professor Emeritus *Donald Levis* from Animal Science, University of Nebraska–Lincoln and Levis Worldwide Swine Consultancy.

Professor *Jay Harmon* from College of Agriculture and Life Sciences, Iowa State University, for permission to redraw and use pictures from several extension and research papers.

General Director *Ralph Addis* from the Department of Primary Industries and Regional Development, State of Western Australia.

To *Terry Loecke, Matthew Liebman, Randy Hilleman, Scott Bauck, Gheata Razvan*, and many other authors mentioned in this book and those who sent us materials for this book.

A brief history of hoop barns

1

In the 1980's Swedish farmers devised an intensive swine production system utilizing deep bedded litter which allowed animals to express their natural behaviors. This system was developed for a relatively harsh mixed continental-oceanic climate receiving in excess of 500 *mm* rainfall per year. The system was also devised in part to allow compliance with restrictive animal welfare regulations (Honeyman and Kent, 1997). This management system's primary premise is the use of group housing in all phases of production. Of particular note is that sows are farrowed in groups where each animal is free to come and go to her individual farrowing box. Farrowing managers do not cross-foster piglets or dock tails. Neither is there any prophylactic antibiotic use allowed. At approximately 10 days post-farrowing (when piglets begin to escape the box) boxes are removed and all sows and litters are allowed to mix. This results in communal nursing (Figs. 1.1 and 1.2).

FIGURE 1.1 Swedish group-farrowing system room during farrowing time.

Group-farrowing system with sows gathering nest material from the straw bedding. The room is 8 × 10 m (80 m^2) with two rows of four farrowing boxes or cubicles for accommodating eight sows. The farrowing boxes are located on the side walls, between feeding area (not visible in foreground) and watering area (back of picture). The sides of the room with farrwowing boxes are darkened to stimulate nest building and farrowing inside the cubicles. Two weeks after farrowing, the boxes will be removed.

Credit: Picture provided by Professor Bo Algers (2019), Swedish University of Agricultural Sciences, Skara,

Sweden.

Alternative Swine Management Systems. https://doi.org/10.1016/B978-0-12-818967-2.00001-0

FIGURE 1.2 Swedish sow group system room during lactation (upper) and after weaning (lower).

Group-farrowing system where the individual farrowing boxes have been removed (upper) and piglets after weaning (lower). At weaning, the sows were moved to a breeding area and the piglets remained in the room. View is from the feeding area end of the room.

Credit: Pictures provided by Professor Bo Algers (2019), Swedish University of Agricultural Sciences, Skara, Sweden.

Piglets are typically weaned at 6 weeks of age by removal of the sows. This management system is designed to ensure animal welfare and high production efficiency (Algers 1997, 2012; Bowman 1993; Honeyman; Kent 1997).

For about 10 years, some American universities have been active in developing alternative swine production systems with the Swedish concept in mind. Various Cooperative Extension programs in the United States and Canada have been promoting these systems which typically utilize deep bedded hoop barns as the primary production facility. As of today, the USA, Canada, Denmark, Sweden, and

FIGURE 1.3 Wisconsin University's Swine hoop barns.

Credit: Hutu (2005), Wisconsin, River Falls, USA.

lately Ukraine, Poland and Romania, have a significant presence of these types of management systems.

While commercial use of deep bedded, tarp covered hoop barns for swine production is a relatively new concept in much of the world, it was practiced as early as the 1970's in Japan where it was known as the *Ishigami system* (Gadd 1993). Canada was also an early adopter of this approach with the advent of the *Biotech system* in 1985. This system used hoop barns with deep bedding and a feeding area located at one end of the shelter. This system has been modified and updated and is now marketed by a number of companies.

Such a swine management system offers significant advantages for Romania and other Eastern European countries as well. The most prominent of these is the low initial investment in structures as compared to that of conventional systems (Onan, 2005). In addition this system requires significantly less energy use for heating, cooling and ventilation (Fig. 1.3). It is also more environmentally friendly with low impact on water quality and reduced odor and other air quality concerns (Algers 1997; Hinrichs; Tranquilla 1998).

With 2000 hoop structures, each with a 200 head capacity and 2.5 turns of production per year, it is possible to produce one million head of finished hogs. In 2004, this was the approximate level of swine production from such systems in Iowa, USA where about 3000 such structures exist, most of which are used for finishing swine (Miller, 2005). This would be an achievable goal for East European countries and Romania as well.

One can conclude therefore, that hoop barns with deep litter bedding can be a cost effective environmentally friendly swine production approach which will hopefully be accepted in East European countries, Romania or in other states as their swine industry develops (Onan et al., 2016).

References

Algers, B., 1997. Cold housing and open housing — effects on swine health, management and production. In: 9th International Congress in Animal Hygiene, 17—21 August, 1997. Helsinki, Finland, vol. 2, pp. 525—527.

Algers, B., 2012. Contemporary issues in farm animal housing and management, swine. In: Jakobsson, C. (Ed.), Ecosystem Health and Sustainable Agriculture. 1. Sustainable Agriculture. The Baltic University Programme, Uppsala University, pp. 329—337.

Bowman, G., 1993. Fitting the farm to the hog. New Farm 15 (6), 35—39.

Gadd, J., 1993. Tunnel housing of pigs. In: Livestock Environment IV, Fourth International Symposium, Michigan. American Society of Agricultural Engineers, pp. 1040—1048.

Hinrichs, C., Tranquilla, J., 1998. A multi-stakeholder view of the potential for hoop structures in Iowa swine production. In: Animal Production Systems and the Environment: an International Conference on Odor, Water Quality, Nutrient Management and Socioeconomic Issues, Proceedings, vol. II, pp. 711—716.

Honeyman, M.S., Kent, D., 1997. First Year Results of a Swedish Deep-Bedded Feeder Pig Production System in Iowa. Swine Res. Rept ASL-R1497. ISU Extension, Ames.

Hutu, I., 2005. Reproducerea intensivă la scroafe. Ed. Mirton, Timişoara.

Miller, L., 2005. Hoop Structure Proves its Worth. Wallace Farmer, January, p. 36.

Onan, G., 2005. The application of hoop structures for growing and finishing swine. Lucr. şt. Med. Vet. Timişoara 38, 342.

Onan, G., Mircu, C., Patras, I., Hutu, I., 2016. Hoop structure for wean to finish pigs: management and gender influence in a summer trial. In: College of Agricultural Science and Veterinary Medicine, Iasi, RO: Veterinary Conference, vol. 59 (4), pp. 388—394.

Location, design, and construction

2

2.1 Hoop barn location and biosecurity

2.1.1 Swine farm location

Swine structures must be sited as required by national regulations for environmental protection and animal biosecurity. Current law requires minimal distances from human housing areas and other places of activity and is designed around a concentric ring concept. The triple rings define the *no access area*, the *fending area*, and the *protection area*. The farm must be contained by a 1.8—2 m high fence. Inside the fenced area where barns are located, the ground must be graded and leveled and covered with grass sod that is routinely mowed. Additional regulations include the following:

- the farmstead will be located in an area with a windbreak and no flood hazard;
- the distance between two consecutive buildings should be 10—20 m (Anon, 2003), the exact distance depending on the construction material's fire resistance and the ventilation requirements of the structure;
- structures should be oriented transverse to the axis of the prevailing wind or sited longitudinally from north to south or northeast to southwest;
- feed and supply storage should be located at the edge of the production area with an access gate located nearby in the boundary fence.

Hoop barns should be located within the protection area in a compact fashion in one or more rows (Figs. 2.1—2.4). The grade around the barns should slope to the north. Typically, the main barn opening will face south.

2.1.2 Swine farm biosecurity

Farm biosecurity must take into consideration the following concerns:

1. **Isolation:** The distances between neighboring production units and between buildings within production units will affect biosecurity. There are a variety of approaches for locating barns within a production unit (Waddilove 1996):
   All housing in one location (Fig. 2.5A): It is a common approach in traditional production units with breeding/farrowing and grow-finish all in one enclosure.

Alternative Swine Management Systems. https://doi.org/10.1016/B978-0-12-818967-2.00002-2

FIGURE 2.1 Hoop barns aligned in a single row.

Credit: Photo by Hilleman, R., 2004. Production cost for bedded livestock operations. In: Hoop Barns & Bedded Systems for Livestock Production Conference, 14 September 2004, Ames, Iowa. With permission of Honeyman, ISU Hoop Conference.

FIGURE 2.2 Hoop barns in a double row alignment.

Credit: Photo by Loza, A., 2004. CJSC Agro-Soyuz Ukraine - hoop barns in the Ukraine. In: Hoop Barns & Bedded Systems for Livestock Production Conference, 14 September 2004, Ames, Iowa, Agro-Soyuz, the Ukraine. With permission of Honeyman, ISU Hoop Conference.

Dual location system (Fig. 2.5B): This is the approach where breeding/farrowing and early growth up to 25—40 kg is at one site, whereas finishing is located in a separate building. Because of the continued emergence of more immunosuppressive viral diseases, it is imperative that animal health maintenance include not only the use of vaccination and medication but also adequate isolation. The epidemiology of most swine diseases dictates that isolation is critical to break the cycle of infection. Some factors that influence disease transmission include the following:

FIGURE 2.3 Five rows of hoop barns in a compact arrangement.

Credit: Photo by Loza, A., 2004. CJSC Agro-Soyuz Ukraine — hoop barns in the Ukraine. In: Hoop Barns & Bedded Systems for Livestock Production Conference, 14 September 2004, Ames, Iowa, Agro-Soyuz, the Ukraine. With permission of Honeyman, ISU Hoop Conference.

FIGURE 2.4 Four rows of hoop barns in a more dispersed arrangement.

Credit: Photo by Hogberg, M., 2004, Context, history, and perspectives. In: Hoop Barns & Bedded Systems for Livestock Production Conference, 14 September 2004, Ames, Iowa, Iowa. With permission of Honeyman, ISU Hoop Conference.

- Young piglets are well protected by passive immunity from their dam's colostrum until the age of 10—14 days. Their own immune development is not complete until about 21—28 days.
- Most disease transmission occurs over close-range nose-to-nose contact.

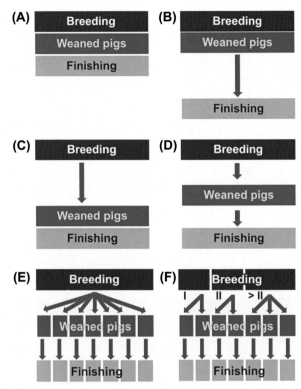

FIGURE 2.5 Location alternatives.

(A—F) from the top to bottom. (A) All housing in one location. (B) Dual location system, (C) Structures sited at two distinct locations. (D & E) Isolated structures sited at three distinct locations. (F) Structures sited at multiple distinct locations depending of parity of the sows (I - young sows at first parity, II - second parity and >II - older parity sows).

Credit: Redraw after Waddilove, J., 1996. Breaking the Disease Cycle-New Patterns in Pig Production, http://www.antecint.co.uk/main/virkons.htm and Harris, D. L. 2000. Multi—site Pig Production, Iowa State Univ, Ames, Iowa Press.

- Most transmission is from older swine to younger animals.
- Swine with disease is the greatest threat to healthy swine.

When all factors are considered, several modern biosecurity models are available as indicated below:

Structures sited at one location: The principle of all-in-all-out is utilized between structures or between isolated units within a structure.

Structures sited at two distinct locations (Fig. 2.5C): Gestation, farrowing, and lactation at one site with grow-finish at a second site (Harris 2000).

Isolated structures sited at three distinct locations (Fig. 2.5D and E): Gestation and farrowing at one site with growing and finishing located at two additional isolated sites. Disease transmission risk at each location is deterred by all-in-all-out management. New animals arrive in batches. Piglets are early weaned and raised in nursery units until weights of 25−40 kg are reached at which point they are moved to finishing barns located at a separate location. Each turnover of animals is accompanied by a complete emptying of all the previous batch of animals from the structure.

Structures sited at multiple distinct locations (Fig. 2.5F): Features of this system are similar to the previously discussed model except that the gestation/farrowing unit is also further subdivided with gilts—whose disease challenges and response are different from those of multiparity sows—handled separately. Piglets from primiparous dams are maintained in separate structures from those of multiparous dams throughout the grow-finish phases as well.

2. **Farm location:** Regulations require that swine sites not be located near community areas, public roads, areas prone to high wind or flooding, or near other animal production units. Aerosols will spread most common pathogens over a maximum of 3 km radius.

3. **All-in-all-out:** This management paradigm must be practiced to prevent disease transmission from one generation of animals to the next, from the oldest animals to the youngest.

4. **Wild animal, bird, and rodent access:** Access by these potential disease vectors must be prevented to the greatest degree possible. It is recommended that the farm perimeter be encircled by 2-m-high tight fence (such as chain link) or high-tension electric fence with close wire spacing. In addition, bird and/or insects nets may be used at all the shelter end walls and inlets.

5. **Visitor access:** Outside human visitors must be kept to a minimum. Access will be provided through a checkpoint with available showers and complete sets of clothing and other protective gear.

6. **Feed:** Feed deliveries will be at specified entry points, and off-loading will take place under covered enclosures.

7. **Operator vigilance:** Managers must maintain a high level of observation to allow for early detection of disease outbreaks or for conditions that might lead to such outbreaks.

2.2 Construction details

Hoop barns are constructed out of light materials and designed to protect the animals inside from environmental extremes such as rain, wind, solar radiation, and within limits, extreme temperatures (Hutu and Onan, 2007).

2.2.1 Sizing and design

Sizing recommendations for hoop barns are quite flexible. The most important consideration is the dimension of openings in the structure relative to the total structure size.

Important and recommended sizing indices (Fig. 2.6) for most swine hoop shelters are as follows:

- Length (L): 20.0–30.0 m
- Width (W): 6.0–11.2 m
- Ratio of W/L: 1/3
- Wall height (h): 1.0 1.8 m
- Height (H): 4.6–5.6 m
- Ratio of H/W: 1/2
- Bay (d): 1.2–1.8 m
- Total area: 80.0–336.0 m^2

The primary supporting members of the structure include wood posts set in concrete-filled holes and steel arches (hoops) fastened to the posts. Posts are typically 15–20 cm square and a minimum of 2.4 m long with half the length set into the ground. Posts should be chemically treated to prevent degradation by rot and termites. The steel arches may be single pieces but more typically are composed of two to four sections, which are bolted together. Diameter of the steel tubing is generally 5–6 cm. Arches must be galvanized to prevent rust.

2.2.2 Structure perimeter location and layout

The long axis of the structure will typically be oriented north and south (Fig. 2.7).

Once the perimeter for the initial structure is laid out, any additional ones can be determined in a similar manner with each structure at least 10 m apart.

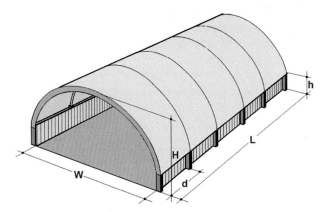

FIGURE 2.6 Main hoop sizing indices.

Swine Experimental Unit, important sizing indices; lenght (L), width (W), height (H), wall heigh (h) and bay (d).

Credit: Ioan Hutu, Horia Cernescu Research Unit Project, 2009, Timisoara, Romania.

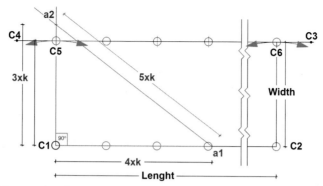

FIGURE 2.7 Layout of perimeter and squaring.

The following is brief procedure for initial layout of the site: (1) locate the access road (east/west) across the south end of the structure; (2) layout one side of the structure, north/south, by setting two posts into the ground at points *C1* and *C2* (to mark barn corners) and tie a string line between them; (3) lay out the shelter width by drawing two arcs on the ground with centers at points *C1* and *C2*; (4) set posts at *C3* and *C4* into the ground along a line drawn tangent to the two arcs previously laid out; (5) to square the corners, compute a constant *k*, where $k = width \times 0.36$, i.e., for a width of 10 *m*, $k = 3.6$ m; (6) lay out the first leg of a right triangle along the *C1/C2* line (*C1* to *a1*); this leg should have length equal to $4k$; (7) mark or cut two strings—one of which will be $3k$ in length to lay out the second leg of the triangle (attach to point *C1*) and one of which will be $5k$ for the hypotenuse of the triangle (attach to point *a1*); (8) the free ends of the two strings will cross at a2. Where the *C1/a2* string crosses, the *C3/C4* line marks the location of the third corner of the structure (point *C5*); (9) on the *C3/C4* axis, measure the structure length from point *C5* to locate point C6, which will be the fourth corner of the shelter. Check for square by measuring the two diagonals to make sure they are equal.

Credit: Adapted after Quick structures, Archseries, Installation Manual, 01/2005, MAN-Q205-R1.PDF,

www.quickstructures.net.

2.2.3 **Positioning of support posts**

After laying out the perimeter as outlined in the previous section, support posts for the four corners of the structure will be in place. Diagonal and lateral measurements taken from the center of the posts should be rechecked to assure that the perimeter is square (Fig. 2.8).

String a line between the corner post centers to serve as a guide for setting the remaining posts. Then mark the locations for the intermediate posts, which can be 1.2−1.8 m apart (the bay length = *a*). Narrower structures will have the longer bay length values, whereas wider structures must have shorter bay lengths.

Choose a final bay length that allows equal distancing (*d* in Fig. 2.8) of the posts along the side. As posts are being set, continue to measure diagonals and bay lengths to ensure that posts line up with their mates on the opposite side of the structure.

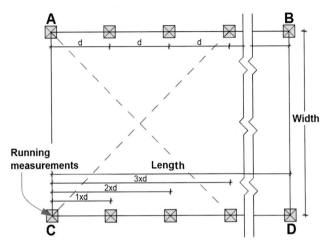

FIGURE 2.8 Post location.

(A–C) structure corners; (D) *d*—bay length (distance in-between each post and its adjacent one); *l*—width; *L*—length (multiple of *d*).

Credit: Adapted after Cover-all, 2005, www.quickstructures.net.

2.2.4 Foundation and setting of posts

The foundation for each post will consist of a gravel bed within each post hole with the remainder of the hole filled with concrete (Fig. 2.9). Corner posts must be set first and vertical alignment checked with a plumb line. Post height is to be set at 1.25–1.8 m (Anon 2003) above grade. It is advised that height of the corner posts be equalized using a transit level. The remaining posts can be leveled by sawing at a height marked by a taught string line stretched between the corners after all posts are set.

Post hole locations can be determined by stretching a string line between the corners and dropping a plumb line from it at the prescribed distances. Once the hole is drilled, the bottom is filled with about 0.1 *m* of gravel and posts are inserted. Each post should have two metal rods (30 *cm* long and 20 *mm* diameter) at 90 degree angles penetrating it. These will assist in vertical positioning of the post during setting and strengthening the post placement. These rods should be located at approximately 0.75 and 0.9 *m* below grade. Make sure to double-check alignment and plumb as concrete is added to the hole.

For heavy metal structures, a reinforced concrete foundation is used and the metal posts are aligned in the same manner as described above (Fig. 2.7).

2.2.5 Installation of arches

Arches, either single piece or multiple pieces bolted together, are bent by the manufacturer at a radius equal to the structure height (*H*). These are set into the mounting

(A)

(B)

FIGURE 2.9 Concrete reinforced foundation and post placement.

Research hoop units during construction. Digging the foundation and placing the posts on reinforced concrete foundations.

Credit: Photo by Ioan Hutu, Horia Cernescu Research Unit, 2012, Timisoara, Romania.

brackets installed in the previous step and fastened with the supplied bolts and nuts (Fig. 2.10). If arches contain multiple pieces, these are connected with joiners, which are pipe pieces whose outer diameter exactly matches the inside diameter of the arch. For various combinations then, the following situations occur:

- two-piece arch has one joiner at the middle (top of the arch);
- three-piece arch has two joiners each located about one-third around the length of the semicircular arch; and
- four-piece arch has three joiners each located about one-fourth around the length of the semicircular arch.

FIGURE 2.10 Assembling and reinforcing the metal framework.

Research units during construction. Mounting of arches to the posts and connecting arches to one another *(left)*. The post for the doors is mounted to the structure ends. Support rods are used between arches to impart rigidity. Note that diagonal bracing rods are doubled between the first two arches at each end *(right)*.

Credit: Photo by Ioan Hutu, Horia Cernescu Research Unit, 2012, Timisoara, Romania.

Each joiner contains a perpendicular rod that is used for fastening spacing pipes that span between adjacent arches. At the ends of the structure, there are additional spacers installed to help support the first arch by attaching it even more securely to the second arch. In areas with potentially heavy snow loads, the hoop frames will need to be made of larger and/or more tubing components (Fig. 2.10).

2.2.6 Floor construction

Regardless of the production phase for which the structure is being used for, the floor area consists of two functional regions—a *feeding area* and a *deep bedded resting area*. The floor material may be entirely concrete or may consist of a combination of a concrete floor in the feeding area and a dirt floor in the resting area. Typically, the resting area is located in the north 2/3's of the structure and the feeding area is the remaining southern 1/3. The same proportion is maintained if resting and feeding areas are located on the sides of the structure (Fig. 2.11).

The floor level at the north end should be 10 cm above the surrounding grade and then slope 1.5% to the south throughout the remainder of the resting area. If the resting area floor is of earthen material, it must be made watertight by including clay[1] in the soil mixture and firmly packing it. Alternatively, buried sheets of

---

[1] It is most desirable for the floor to be concrete, especially if the structure is being used for farrowing. There is a relatively high risk of ground water contamination using earthen floors. Recent measurements have shown increased ground concentration of $N-NO_3$ at a depth of 30 cm. Increases have ranged from 1 to 5.5 ppm higher after the first turn of finishing swine in a hoop structure and then up to a 14.5 ppm increase following the second group of finishing hogs. At a depth of 1.2 m, the increases were from 1 to 3 ppm following the second cycle of hogs (Richard et al. 1998). We therefore advise that floors must be watertight and therefore likely concrete will be necessary.

FIGURE 2.11 Floor construction—resting and feeding areas.

Swine Experimental Unit during construction and before installing the feeding equipment and pens. Pouring concrete to build resting and feeding areas (*upper*) . Making steps to access the feed area (lower).

Credit: Photo by Ioan Hutu, Horia Cernescu Research Unit, 2012 & 2013, Timisoara, Romania.

polyethylene may provide a water barrier. Water impermeability is necessary because even though deep bedded, there is the possibility of leaching of manure nutrients out of the bedded area and into ground water. In the United States, the standard regulation is to achieve a water permeability of less than 0.1 mm/day (Kruger et al. 2006).

The feeding area is necessarily composed of concrete. This area is generally at a grade 0.4 m higher than the grade established at the south end of the structure. The feeding floor should slope 1% to the south to allow water from fountains to flow out of the building rather than into the bedded area (Gegner, 2005). Steps 20−25 cm wide and 15−20 cm high should be provided at the juncture of the feeding and resting areas to allow easy passage between the two for young piglets.

2.2.7 Installation of sidewalls

Sidewalls are constructed of tongue-and-groove lumber fastened to the inside of the posts. These sidewalls help to reinforce the structure by stabilizing the posts to which the arches are mounted. If the posts are wood, the step-by-step procedure is as follows:

- After posts are set, attach one 5×10 or 5×15 cm board to the outside of the posts flush with the top of the posts and fasten with nails 10–12 cm long.
- Attach five to eight tongue-and-groove boards to the inside of the posts. These may be 5×15 to 5×25 cm boards and should be in lengths equal to even multiples of the bay length (d). The entire height of the wall from the ground to the top of the posts should be boarded. Fasten boards with nails 10–12 cm long.
- Mount the arch support brackets to the center of the top of each post with lag bolts.
- Mount the tarp tensioning winches on the outside of each post at a distance 60 cm from the top.

If an entirely metal structure is used, wall assembly follows basically the same steps except that self-tapping screws or similar fasteners must be used. In either case, there should be a minimum 10 cm horizontal gap between the top of the wall and the bottom of the tarp (Fig. 2.12).

2.2.8 Construction of air deflection panels

The purpose of deflection panels is to direct air flow entering the open space at the top of the posts located between the inner sidewall and the outer board. The width of this space is equal to the post dimension (15 cm). It serves as a natural air inlet, but this air must be directed upward such that it does not cause a draft for animals housed in the building.

The deflection panel is typically made from 1.25- to 2.0-cm-thick plywood, which is 60 cm wide. The bottom edge is fastened to the sidewall using screws. A mounting strip made of a 5×10 cm board is attached to the inner side of the arches with U-bolts. The top edge of the deflection panel is then fastened to this strip using screws (Fig. 2.13).

The space between the mounting strip and the underside of the tarp covering the hoop structure then constitutes the available opening for air entrance. Some individuals construct the deflection panel with hinges on the bottom edge and hooks with various settings at the top so that the air inlet slot size can be adjusted to control air flow more precisely. Generally speaking, this is not a necessary requirement for good ventilation, however.

2.2.9 Tarp installation

The arches are covered with a tarpaulin, which may be white, silver, or light blue or green (Figs. 2.15–2.17). In addition, tarps may be opaque or semitransparent and are made of silicon and polyurethane material, which has been made ultraviolet light resistant to prevent solar degradation over time. Tarps are typically one piece with pockets sewn into each longitudinal edge into which steel pipes are inserted for the entire length.

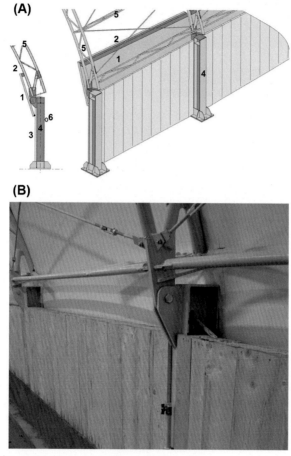

FIGURE 2.12 Side wall and air deflection panel assembly.

(A) Swine Experimental Unit project, 2010: 1. deflection panel; 2. slat at upper fastening point of panel; 3. side wall; 4. post; 5. metal double arches; 6. pipe inserted into sleeve at bottom of tarp for tensioning and positioning of the tarp.
(B) Picture: Wood wall without deflection panels. Note the space between the wall and tarp. After air flow measurements were made within the finished structure, it was determined that deflection panels were not necessary.

Credit: Photo and drawing by Ioan Hutu, Swine Experimental Units, Horia Cernescu Research Unit 2010 & 2012, Timisoara, Romania.

Usually, the tarp is pulled over the arches by attaching ropes to the pipe along one side and pulling the tarp up and over the arches or as in Fig. 2.13. Once in place, the tarp is tensioned using the winches attached to every first or every second post.

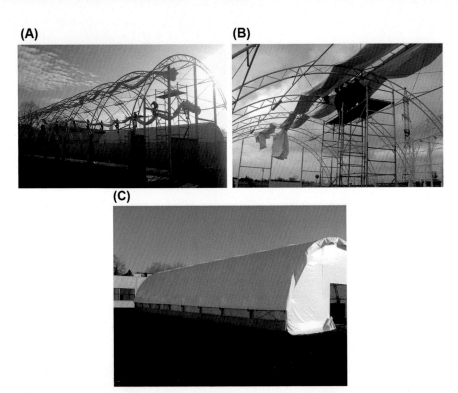

FIGURE 2.13 Covering and tensioning of the tarpaulin on hoop arches.

(A) Swine Experimental Unit during construction. Raising the tarpaulin halves over the metal arches of the structure. (B) Attaching the tarpaulin halves to create the ventilation slot. (C) View of the structure after tensioning tarpaulins on the side walls using winches. The edge of the tarpaulin is not yet tensioned and fastened. The end walls can be completely removed by folding from the top down. The tarpaulin can also be removed from the doorway.

Credit: Photo by Ioan Hutu, Horia Cernescu Research Unit, 2012, Timisoara, Romania.

FIGURE 2.14 South end of two hoop structures.

Credit: Photo by Maltman, J., 2004. A review of pig production perspective in hoop shelters on the Canadian prairies, Hoop Barns & Bedded Systems for Livestock Production Conference, 14 September 2004, Ames, Iowa, Manitoba, Canada. With permission from Honeyman, ISU Hoop Conference.

FIGURE 2.15 Tarp fastening and tightening over framework.

The tarp covering is tightened by putting tension on the canvass straps with winches as shown above. The canvass straps are wrapped around the pipes inserted into the continuous pocket along the length of the tarp. Once tensioned, the tarp will cover the top of the sidewalls and posts and extend down over the outer edge of the sidewalls a few centimeters.

Credit: Photo by Gary Onan, 2007, River Falls, Wisconsin.

A webbed strap is inserted around the pipe in the pocket at the tarp's edge, and the winch is used to pull the webbed strap tight. The tarp is tensioned longitudinally by using a long continuous cord supplied by the manufacturer.

The cord is alternately laced through eyelets sewn into the underside of the tarp near its end and then around the first arch, thus pulling lengthwise on the tarp (Fig. 2.16). The free edge of the tarp at the ends is also tensioned by using a cord and winches. A long cord is inserted into a continuous pocket sewn along this edge, and the cord is tensioned at each end using small winches attached to the corner posts of the structure (Fig. 2.14).

For farrowing facilities, multiple-layered tarps are available for better insulating ability. A pocket of air between adjacent layers serves this purpose. The air is introduced using a large compressor creating an air pocket 5—10 cm thick. In addition, the outer layer will be transparent allowing solar radiation to heat these air pockets. If water or snow collects on the tarps, that is a good indicator of insufficient tautness. If properly tensioned and with the animal heat from inside, the tarp should allow snow to slide off to the sides of the structure.

2.2.10 Shelter endwall construction

The lower portion of the endwalls will be constructed of tongue-and-groove boards just as the sidewalls are and should be at a comparable height (Fig. 2.12). The arched portion of the endwalls is typically covered by canvas material similar to the main tarp over the arches. Manufacturers will supply ready-cut pieces that will properly fit the radius of the arches. These pieces are strapped into position (Fig. 2.16) to a moderate tension.

These canvas pieces are typically removed during warm weather seasons to allow more airflow for ventilation and are reinstalled during cold weather seasons to help maintain warmth within the structure. It is imperative that even in cold weather the uppermost portion of the endwall be left open to allow for minimum adequate ventilation (Figs. 2.16 and 2.18).

FIGURE 2.16 Structure ends—details.

South end, tarp is covering sides of door and top is open for ventilation (left-up). North end, with tarp at upper end hanging open for ventilation (right-up). North end, note the cord looped around the outside arch and tensioning the tarp lengthwise (left-down). South end as viewed from inside with self-feeders located in feeding area (right-down).

Credit: Photo by Gary Onan, 2007, River Falls, Wisconsin.

FIGURE 2.17 Outside view after construction.

At the end of the construction phase for two structures (upper) was covered with 40 cm of snow in an unusually hard winter in Europe. The double-tube arches supported the snow load for several days. Part of the snow slid off the sides during the following days (lower).

Credit: Photo by Ioan Hutu, Swine Experimental Unit, Horia Cernescu Research Unit, 2012 & 2019,

Timisoara, Romania.

FIGURE 2.18 Inside of the structure—resting and feeding areas.

Inside of structure before installing the equipment (left). Inside of the structure after equipment installation—the feeding area (left of picture) and the resting area (right of the picture) are separated by a step. The top ventilation slot in the tarp will be permanently open but covered with a ridge cap to keep out rain.

Credit: Photo by Ioan Hutu, Swine Experimental Unit, after installing the feeding equipment and pens, 2013,

Timisoara, Romania.

References

Anon, 2003. Notes from Farmer Round-Table Discussion Around the State of Minnesota. University of Minnesota, pp. 10. www.regionalpartnerships.umn.edu/westcentral/%20projects-ASPRTP-01.pdf.

Gegner, L., 2005. Hooped Shelters for Hogs. ATTRA. www.attra.ncat.org/attra-pub/hooped.html p.1-16.

Harris, D.L., 2000. Multi—Site Pig Production. Iowa State Univ. Press, Ames, Iowa.

Hilleman, R., 2004. Production cost for bedded livestock operations. In: Hoop Barns & Bedded Systems for Livestock Production Conference, 14 September 2004, Ames, Iowa.

Hogberg, M., 2004. Context, history, and perspectives. In: Hoop Barns & Bedded Systems for Livestock Production Conference, 14 September 2004, Ames, Iowa.

Hutu, I., Onan, G.,W., 2007. In: Mirton (Ed.), Tehnologii Alternative Pentru Cresterea Porcilor. Timisoara.

Kruger, I., Taylor, G., Roese, G., Payne, H., 2006. Deep-litter Housing for Pigs. In: PRIME-FACT, vol. 68. NSW Department of Primary Industries, State of New South Wales. http://www.dpi.nsw.gov.au.

Loza, A., 2004. CJSC Agro-Soyuz Ukraine — hoop barns in the Ukraine. In: Hoop Barns & Bedded Systems for Livestock Production Conference, 14 September 2004, Ames, Iowa.

Maltman, J., 2004. A review of pig production perspective in hoop shelters on the Canadian prairies. In: Hoop Barns & Bedded Systems for Livestock Production Conference, 14 September 2004, Ames, Iowa.

Richard, T.L., Harmon, J., Honeyman, M., Creswell, J., 1998. Hoop structure bedding use, labor, bedding pack temperature, manure nutrient content, and nitrogen leaching potential, ASLR1499. In: 1997 ISU Swine Research Report, AS-638. Dept, of Animal Science, Iowa State University, Ames, IA, pp. 71—75.

Waddilove, J., 1996. Breaking the Disease Cycle-New Patterns in Pig Production. http://www.antecint.co.uk/main/virkons.htm.

Environmental control and structure management

3.1 Ventilation

Hoop barns require no artificial ventilation (Figs. 3.1 and 3.2). Natural flow through the structure serves to refresh the air and remove moisture accumulation.

These air currents are of two types:

- longitudinal
- ascending transverse

The tunnel effect created by the long narrow shape of the building establishes the longitudinal flow pattern. During warm seasons, when the ends of the structure are open, the longitudinal flow is a major contributor to air replacement. During cold seasons, longitudinal flow is diminished owing to the closing of the endwalls, but it still contributes to ventilation as warmer humid air rises toward the structure ceiling and is removed through the more open south endwall (Richard et al. 1998).

FIGURE 3.1 Ventilation of barn during winter at north endwall.

Credit: Photo by Gary Onan, 2007, River Falls, Wisconsin.

Alternative Swine Management Systems. https://doi.org/10.1016/B978-0-12-818967-2.00003-4

FIGURE 3.2 Ventilation details.

Views showing deflection panels and open north door *(upper left)*; removal of tarp at shelter ends over summer *(middle left)* and flap inlets *(lower left)*. Central superior ventilation opening *(lower and upper right)*.

Credit: Photo upper left: Maltman, J., 2004. Manitoba agriculture, food and rural initiative, hoop barns in Canada, In: Hoop Barns & Bedded Systems for Livestock Production Conference, 14 September 2004, Ames, Iowa, Canada. With permission of Honeyman, ISU Hoop Conference. Photo by Ioan Hutu, 2004, Wisconsin, River Falls, USA (middle left) and 2008, Experimental Units, Timisoara, Romania (lower left and right).

The rising humid air creates cross-currents as outside air is admitted through the slots at the top of the longitudinal sidewalls. Control of natural air flow within the structure is therefore critical to optimize ventilation. It is recommended that the following approaches be used:

- During winter, direction of longitudinal currents is primarily from north to south through the structure. The north (cold) end should be almost entirely closed. Most incoming air is therefore through the longitudinal slots. This air flow is directed upward by the deflection panels and by the natural air flow created by the rising of warmer air created by the presence of the livestock at floor level. The primary exhaust point for this air will be the south end wall, which will be more open than the north one. If the inlet slots along the sidewalls are under-sized or closed, significant moisture buildup and mold accumulation can result.
- During summer, the endwalls are completely opened to maximize air movement through the structure. Sidewall slots are kept entirely open, and deflection panels may even be removed (Honeyman. 1997, 1998).

At the floor level, an air flow rate of 0.15 m/s is considered comfortable for swine. At a flow rate of 0.5 m/s, swine will feel 3°C cooler than the actual temperature (wind chill effect), and at a flow rate of 1.0 m/s, they will feel 6−7°C cooler than ambient temperature (www.thepigsite.com).

In hoop barns used as farrowing units with farrowing boxes set along the sidewalls, there may be a disturbance of the natural air flow patterns owing to the solid walls of the boxes. These will create turbulent flows that will slow the overall air movement and make effective control of air flow much more difficult to achieve. Therefore, it is recommended that during winter in particular, boxes be placed 5−10 cm from the sidewall to allow a passageway for the natural air currents within the structure (Rossiter and Honeyman, 1997).

3.2 Temperature and humidity control

Temperature within hoop barns follows the outdoor ambient temperature and is dependent on the season, day, and hour. Hoop barns are in fact cold barns, and as a rule, no heaters are used. Our research (Onan et al., 2016) and a number of experts (Jannacsh 1996; Harmon and Xin, 1997; Honeyman and Harmon, 2003) indicate that during cold months, the internal shelter temperature is 3−6°C higher than outside, and during warm months, it is 1−2°C higher. The external to internal temperature gradient depends on the following factors:

- Structure length: Air exchange is typically inadequate in structures over 30 m long
- Number of tarp layers and tarp color: dark colors absorb more heat
- Time of day: temperature in the barn follows ambient temperatures and so is naturally higher in late afternoon and is maintained at that higher level for a longer period into the evening and night time.

During summer, barn temperatures can be maintained within permissible limits by:

- fully opening endwalls and maximizing open space along sidewall slots;
- use of fans to speed up natural air currents (Messenger 1996);
- use of fogging or misting devices or in the case of farrowing units with individual boxes by a water drip onto the sow's head;
- use of a row of tall deciduous trees for shading;
- use of a thinner layer of bedding so swine can get closer to direct contact with the ground and thus lose heat via conduction to the cooler earth; pigs will also lie spaced farther from one another which facilitates cooling as well.

Regardless of the season, if the structure is used for farrowing, temperature and humidity control are more critical. If summertime temperatures get too high inside a farrowing unit, a number of undesirable events are likely to occur such as increased number of stillborns, increased crowding, and splashing around watering areas to create wet spots for cooling, reduced feed intake, and avoidance of entering farrowing boxes (Payne 2004). During winter, barn temperatures can be maintained within permissible limits by:

- lowering air flow speed by closing the north endwall and partially closing the south endwall and by restricting the slot width along the sidewalls;
- decreasing heat loss by covering with an insulating cover; such a cover may contain two to seven layers of tarp material, some of which may be composed of insulating materials or may be arranged such that there are pockets of trapped air between the tarp layers;
- use of deeper bedding, which is more frequently replenished;
- use of heaters—either large general heaters or localized area heaters (Fig. 3.3).

Area heating may be accomplished by using infrared heat lamps mounted above farrowing boxes or infrared heaters mounted on the walls of gestation units. For farrowing boxes, only one lamp is required per box. Energy consumption by such a lamp will be about 14—21 kWh during the first week following farrowing and 10—15 kWh during the second week (Stuhec et al., 2002).

Weaned piglets can be housed in heated pens (3.0 × 1.2 m) with walls about 1.0 m high and the front pen panel open to allow for air flow. Typically, heat can be provided by four evenly spaced heat lamps hanging above each pen with 250 W bulbs in each.

General heating for farrowing or nursery units may be provided by radiant tube gas or electric heaters mounted onto the structure's framework. Quite elaborate and effective heating systems can be designed by utilizing smaller hoops within the main hoop. This allows for creating warmer confined spaces for piglets while maintaining cooler areas for sows (Figs. 5.16 and 5.17).

Ideal relative humidity inside hoop structures should be 60%—70%. The combination of relative humidity and temperature is the more important factor, however, as the effect of high temperature is much more severe if humidity is also high. Fig. 3.4

FIGURE 3.3 Heating devices used in hoop structures.

Gas heaters for general warming—*upper left*. Boxes with heat lamps for local warming—*upper right*. Infrared radiant panels, normal and infrared images—*lower*.

Credit: Photo: https://www.leopold.iastate.edu, with permission of Honeyman (upper left); and photo by Ioan Hutu, 2018, Experimental Units, Timisoara, Romania (upper right and lower).

illustrates the relative levels of intervention required for various combinations of temperature and humidity.

Extremes of humidity should be avoided if at all possible. At very high humidity, the ability of animals to effectively disperse excess body heat is extremely impaired. At humidity levels below 50%, on the other hand, breathing may be impaired due to dust buildup.

Interventions as required by Fig. 3.4 may be accomplished by adjusting air flow rates through endwalls and side slots, by adjusting bedding levels, and to some degree by adjusting feeding schedules such as feeding at nighttime.

3.3 Feed and water delivery systems

Feed and water equipment should meet the following requirements:

- Allow for easy access so there is minimum competition for feeding or drinking space.
- Prevent contamination by anything potentially harmful if consumed (Figs. 3.5–3.9).

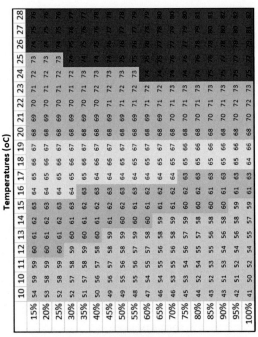

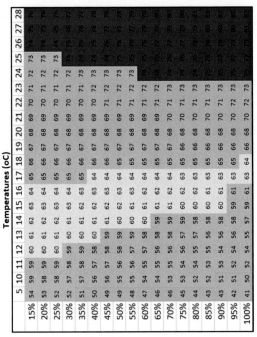

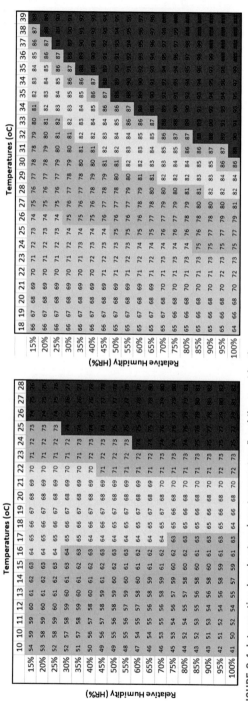

FIGURE 3.4 Intervention levels at various temperature/humidity combinations.

The charts of heat stress for gestating sows (upper left), finish hogs (upper right), lactating sows (lower left) and piglets (lower right). The chart for heat stress combines the effects of both temperature and relative humidity to provide classification as comfort (yellow area), alert (orange area), danger zone at lower temperature (blue area) and danger zone at higher temperature (red in the chart) for several categories of swine based on temperature (°C) and relative humidity (RH%) index, using the equation [I.T.U = 1.8*T+32-(1-RH%) (T−20)] of Dinu 1978.

Credit: Hutu (2019).

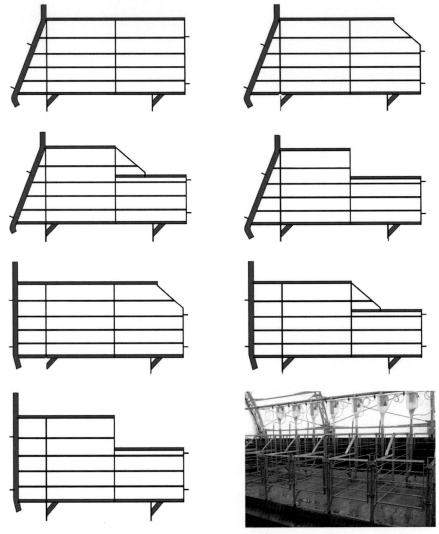

FIGURE 3.5 Types of individual feeding stalls for sows.[1]

Slope front straight back (upper left), slope front slope back (upper right), slope front modified AI back (second row, left), slope front AI back (second row, right), straight front slope back (third row, left), straight front modified AI back (third row, right), straight front AI back (lower left), Pictures of individual feeding stalls for sows in Swine Experimental Units, Timisoara, Romania (lower right).

Credit: www.hogslat.com and photo by Ioan Hutu, 2019, Horia Cernescu Research Unit, Timisoara, Romania.

- Should be functional under any environmental condition (especially extreme temperatures).
- Present no physical danger to swine.

---

[1] Reference to products in this book is not intended to be an endorsement to the exclusion of others, which may be similar. The owners of farms using such products assume responsibility for their use in accordance with current directions of the manufacturer.

FIGURE 3.6 Feeding stalls that may be individually locked.

Sows may be locked inside stalls by means of individual gates operated by levers accessible from main alley. Center of photo: location of frost proof water fountain.

Credit: Harmon, J.D., Honeyman, M.S., Kliebenstein, J.B., Richard, T., Zulovich, J.M., 2004. Hoop barns for Gestating Swine. In: Rev. Ed. AED44. MidWest Plan Service. Iowa State Univ., Ames, IA. p.20, Midwest Plan Service, Iowa State University, USA.

Feed and water are typically made available in certain areas of the structure designed specifically for that purpose. A variety of equipment styles and arrangements may be used. The specific setup will usually depend on the production stage for which the building is designed.

3.3.1 Feed delivery

Mechanized feed delivery systems are advisable for hoop structures used for housing gestating sows. Typically, no automatic delivery system is used for other types of units, but attention should be paid to make sure feeders are properly matched to pig weight, and the filling schedules are adequate to keep feeders filled with the proper ration. Automatic systems may be either the auger type or the cable type. Cable types are systems controlled by computerized electronics. Feed movement is via a continuous cable running in a tube. At intervals along the cable, paddles are attached, which pull the feed through the tube as the cable is moved along by an electric motor–driven transmission. Systems may be single or double branched and require 2.2–3 kW motors for power and can typically move 1.6–3 t/hour. With either the auger or cable systems, automatic fill switches or stop baffles can be used to limit the amount of feed distributed into each container.

FIGURE 3.7 Self-feeder and animal toy in a hoop structure.

Upper—Feeding area with feeders (right and left in the picture) and wood and chain animal toy (right). Lower—Rectangular self-feeder *(left)*, round self-feeder *(middle)*, boar self-feeder *(right)*.

Credit: Photo by Ioan Hutu 2016, Experimental Units, Timisoara, Romania.

3.3.2 Feeding equipment

Feeding stalls (Figs. 3.5—3.7) can be used to control individual feed amounts for each sow in gestation and breeding units (Brumm, 1999).

The stalls are typically 50—60 cm (maximum 75 cm) wide and 215—230 cm long. Each stall has a feed pan or trough and perhaps a water nipple or bowl at

FIGURE 3.8 Electronic feeder in a hoop structure.

Credit: Harmon, J.D., Honeyman, M.S., Kliebenstein, J.B., Richard, T., Zulovich, J.M., 2004. Hoop Barns for Gestating Swine. Rev. Ed. AED44. MidWest Plan Service. Iowa State Univ., Ames, IA. p.20, Midwest Plan Service, Iowa State University, USA.

the front (although it is not common to provide water in the stalls for sows in hoop structures as they are free to move about within the bedded area during most of the day and have access to a communal water fountain. There are gates at both the front and back ends of the stall so that sows may be locked inside the stall when required for management procedures. Front gates may be opened to allow sows into the feed alley for moving to different locations. Stalls may be constructed such that individual stalls may be closed or they may be equipped with gang-locking systems that close the gates on all stalls at once after sows have entered.

Flooring in stalls is typically constructed of rough concrete with a slope of 1.5% front to back and a slope of 0.5%−1.0% from end to end for the feeding trough area.

The feeding trough is typically constructed of stainless steel, smooth concrete, or plastic panels. Sizes of feeding troughs will vary depending on the exact construction of the stall. Typically, they are 18−22 cm deep at the sow's side and 5−6 cm deep at the feed alley side. The important point is that each sow has her own individual feeding area. This allows for individually tailored amounts of feed for each sow daily and reduces aggression during feeding time. The ability to temporarily lock sows into the stalls allows for easy performance of management activities such as vaccine or medication injections, insemination, pregnancy detection, and easy movement and relocation of the animals (Honeyman and Kent, 2000).

Electronic feeding stations may also be used in gestation units (Fig. 3.8) but require significant initial capital expense. Some experts, however, endorse their use (Olsson et al. 1992).

FIGURE 3.9 Feeders and water fountains in a hoop barn.

Round self-feeder with lids—view from inside structure (*upper left*) and from outside structure (*upper right*). Heated water fountains with lids (*lower*) fastened to the concrete floor at barn corners in feeding area.

Credit: Photo by Gary Onan, 2007, River Falls, Wisconsin, USA.

Drop feeders may be used for filling the feed troughs in the gestation stalls. These systems consist of a small bin above each stall, which is filled by an automatic delivery system. Each small bin contains an adjustable baffle that controls the extent to which it fills. The drop feeder bins may be triggered to empty manually but are typically triggered by an electronic timing device that opens all bins simultaneously, which drops the feed into the trough. A particular design of these feeders allows the feed to drop slowly into the trough at a rate of 0.25–0.5 parts of the total amount per minute. This slow trickle of feed causes sows to stay in their feeding stall waiting for more feed to flow and thus reduces aggression between sows of differing hierarchical standing during feeding time. For nursing sows or grow-finish swine where ad libitum feed availability is desired, typical round cone-bottomed self-feeders may be utilized (Figs. 3.7 and 3.9).

3.3.3 Watering equipment

Water delivery installations should meet two basic requirements:

- Ensure enough water is supplied at adequate flow rates (Table 3.1).
- Prevent freezing of water in cold weather (Fig. 3.9).

Water supply and delivery system, including the well and supply lines, must provide adequate pressure and flow rate to meet the requirements as set forth in Table 3.1. Requirements depend on the stage of production and season of the year (DEFRA and Carr, 1988). Water:feed ratios ranging from 1.78:1 to 2.79:1 for pigs weighing from 20 to 125 kg and fed dry feed ad libitum have been reported by Brumm et al. (2000). Table 3.1 values are typical for summertime needs.

Water fountains are located on the concrete feeding area of the hoop barn (Fig. 3.10) and should be set at the proper height for the size of pigs in the structure (Honeyman and Harmon, 2003). Oftentimes, watering devices that have adjustable heights may be used (such a nipple waterers) and should be set at the proper level as indicated by Table 3.2. Typically for animals receiving ad libitum feed, there should be one water fountain for each 20 animals. For management systems utilizing restricted feeding, one fountain per 30 animals is adequate.

Water medications can be an effective means for the treatment and prevention of disease. During periods with air temperatures above the freezing point, a water medication system such as that in Fig. 3.11 can be used.

3.4 Lighting and noise

The tarps covering hoop structures can be transparent, and therefore, no artificial lighting is required unless it is necessary to perform management procedures during

Table 3.1 Daily water consumption (at thermoneutral temperature), water flow rate, and nipple drinker height for swine of various ages.

Age	Height of nipple drinker at 90° or 135° angle (cm)	Water flow at pressure of 1 PSI mL/minute	Daily water consumption liters
Nursing piglets	15–25	300	Up to 1.5
Weaned piglets, 7–30 kg	25–45	500–800	1.5–2.0
Growing pigs, 30–60 kg	45–60	800–1000	2.0–5.0
Finishing hogs, 60–110 kg	60–75	1000–1500	5.0–6.0
Sows and gilts in breeding or gestation	75–90	1800–2200	5.0–10.0
Nursing sows	75–90	1800–2200	15–30
Boars	75–95	1800–2200	5.0–25

Horia Cernescu Experimental Unit consumption during trials during 2014–2018.

FIGURE 3.10 Location of water fountains and rectangular self-feeders.

Credit: Photo: Payne, H., 2004. Deep-litter housing systems an Australian perspective, In: Hoop Barns & Bedded Systems for Livestock Production Conference, 14 September 2004, Ames, Iowa, Australia. With permission of Honeyman, ISU Hoop Conference.

Table 3.2 Typical quantities of bedding required for hoop housing of swine.

Bedding material	kg/year/sow	kg/hog/cycle		
		Average	Summer	Winter
Shredded corn stalks	305–455	90	60–70	90–115
Corn cobs	365–545	110	70–80	110–140
Barley straw	365–545	110	70–80	110–140
Oat straw	270–410	85	50–60	80–100
Wheat straw	340–510	100	65–80	100–130
Hard wood sawdust	510–760	150	95–115	150–190
Pine/fir sawdust	305–455	90	60–70	90–115
Hard wood shavings	510–760	150	95–115	150–190
Pine/fir shavings	380–510	115	70–85	115–145

After Harmon, J.D., Honeyman, M.S., Kliebenstein, J.B., Richard, T., Zulovich, J.M. 2004. Hoop barns for Gestating Swine. In: Rev. Ed. AED44. MidWest Plan Service. Iowa State Univ, Ames, IA. p.20, MidWest Plan Service, Iowa State University.

nighttime hours. Natural solar lighting may be direct or reflected, and its intensity depends on the season of the year, the degree of haziness, the transparency of the tarp, and the amount of light reflection off interior surfaces.

If artificial lighting is required owing to management schedules, a typical intensity level would be 40 lumen/m^2. Artificial lighting is usually directed at the floor (Fig. 3.12) as wooden sidewalls do not reflect light very well. Four optimum production, photoperiod should be a minimum of 8 h at a minimum of 40 lux at swine level.

FIGURE 3.11 Water medication system.

Water dosimeter arrangement at the Swine Experimental Unit. The apparatus consists of an input hose (yellow), which feeds through a check valve into a water meter, pressure gauge, and filter before the valve manifold. In the picture, the valves are set to bypass the dosimeter. If medication is desired, the middle horizontal valve would be closed, and the two vertical valves opened. The manifold has a second pressure gauge at its outlet after which water is routed through a final filter (green) and water meter and into the supply line for the nipple drinker distribution system. The entire apparatus can be bypassed by connecting the yellow hose directly to the green filter.

Credit:Photo by Ioan Hutu 2018, Horia Cernescu Research Unit, Timisoara, Romania.

FIGURE 3.12 Lighting system.

Fluorescent tubes (left) and regular bulbs (right).

Credit: Photo by Ioan Hutu 2018, Experimental Units, Timisoara, Romania.

Light fixtures may be either a bulb or fluorescent tube type. Bulb life will be about 1000 h and produce about 15.5 lumen/watt. Fluorescent tubes typically last 20–25 times longer and are 3–4.25 times brighter per *watt*. As the source of light is of no consequence to production, the most economical system is usually chosen.[2] Because of the presence of occasional high humidity and the fire hazards associated with livestock housing, all electrical installations should be installed with protective measures such as conduit. All installations must meet local electrical codes.

For all phases of production, maximum noise levels should be below 85 *dB*. It is especially important that noise level be controlled in nursery areas. Noise levels can be controlled by:

- proper maintenance of machinery and equipment;
- increasing distance between noise sources and swine;
- establishing barriers between noise source and swine;
- controlling timing and frequency of noise creating farm activities (Metcalfe 1999);
- maintaining the proper animal density in pens;
- minimizing delivery time for feeding and heat detection in breeding/gestation units.

3.5 Deep litter bedding

Deep litter bedding is typically maintained at a depth of 30–45 cm. A variety of materials may be used for bedding. The bedding material performs the following functions:

- provides an insulating medium to ensure thermal comfort for the animals;
- maintains animals in a dry, clean environment; and
- absorbs animal waste and other moisture, which reduces odors.

In addition, the bedding material should be relatively free of dust and not harbor any potentially harmful microbes. In a hoop barn where the bedding material is properly applied and maintained, animal comfort will be high and environmental conditions will be very acceptable with low odor levels and low humidity (Houghton 1997). Bedding is typically brought into the structure through gates in the north end. Except for wet areas that require immediate addition of new material, bales of bedding are typically left intact and the pigs are allowed to tear them apart and spread the material (Figs. 3.13 and 3.14). This provides an opportunity for animals to perform stimulating physical activity and helps prevent such behaviors as tail biting. The swine will determine where their dunging areas will be within the bedded

---

[2] Regular bulbs have shorter service lives and only convert 10% of the energy they use into light, the remainder being wasted as heat.

FIGURE 3.13 Straw bales for bedding in resting area.

Credit: Photo by Gary Onan, 2005, River Falls, Wisconsin, USA.

FIGURE 3.14 Bedding bales arranged to reduce drafts.

Credit: Photo: Brumm M.C., Dahlquist J.M. and Heemstra J.M., 2000. Impact of feeders and drinker devices on pig performance, water use, and manure volume, Swine Health Prod. 8, 51—57.

area (Fig. 3.15). It is advisable to add fresh bedding to dunging areas frequently so that moisture is absorbed rapidly and fermentation is begun. The fermentation is an important source of heat for the swine during winter months, but it is, however, not desirable during summer.

The quantity of bedding required depends on the season of the year and the production stage of the animals housed in the unit. Table 3.2 gives estimates for the amount needed for different production stages. As only 75% of the floor area is

FIGURE 3.15 Dunging areas in a structure for finishing hogs.

Dunging areas can be noticed along periphery of barn's north end (*left*) and at the junction of the feeding and rest areas (*right*).

Credit: Photo by Gary Onan, 2005, River Falls, Wisconsin, USA.

deep bedded, grow/finish swine will therefore typically use on average about 125 kg/m^2 of resting area during each cycle in the hoop. Groups raised in winter will need about 145 kg/m^2, but groups raised in summer require only about 88 kg/m^2.

The temperature of the deep bedded pack will vary depending on season and the type of bedding material used. For example, straw bedding will reach a temperature as high as 40°C at a depth of 15 cm even in winter (Honeyman et al. 2001). Chopped corn stalks, however, do not produce temperatures as high, reaching only about 30°C at a 20 cm depth during winter (Honeyman et al. 2002).

The moisture level of the bedding pack will also vary with season and with the production stage of animals housed in the hoop barn. Generally speaking, bedding in gestation units will be dryer than that in grow/finish units. Moisture levels will tend to be higher during cold seasons and lower during warm seasons—typically around 70%–75% in winter and 30%–60% during summer. In wintertime, with relatively low air flow through the structure, excessive moisture in the bedding can result in steam rising from the pack, and therefore, more frequent addition of bedding material is required (Rossiter and Honeyman, 1997).

Different bedding materials also exhibit varying levels of absorbency. Barley straw, for example, has very high absorbency (2.86 times its weigh). Other materials are as follows: 2.70 for chopped corn stalks; 2.08 for shredded paper; 1.97 for triticale straw; and 1.15 for wood shavings (Voyles and Honeyman, 2006).

Emission rate and concentration of manure gases in deep bedded hoops is comparable with that of conventional housing systems (Jacobson et al., 2004).

Ammonia concentration in hoop structures is 8–9 ppm during winter (compared with 8.5 ppm in conventional structures) and 5–6 ppm during summer (compared to 5.1 for conventional). Emission rate of ammonia is higher from bedded packs than from conventional manure storage both in winter (0.39 *mg/s/* animal vs. 0.02) and summer (0.43 mg/s/ animal vs. 0.06). Even though emission is higher, concentration is about the same as a consequence of the better ventilation in hoop structures.

H_2S concentration is lower in hoop barns than in conventional systems both in winter (9 ppb vs. 149 ppb) and summer (10 ppb vs. 97 ppb). Emission rate of H_2S is comparable with that of conventional systems during all seasons (0.39 µg/s/animal in winter and 0.55 µg/s/animal in summer).

A *carbon/nitrogen* ratio of *35:1* should be maintained within the bedded pack to minimize ammonia emission (Halverson et al., 1997). This can be accomplished by prompt addition of fresh bedding as needed.

3.6 Animal handling

Over the course of a production cycle, animals will necessarily be handled to accomplish movement between production sites, sorting into lots, or administering veterinary procedures (Fig. 3.17).

A washing station is desirable for sow units to have a convenient area for cleaning sows before movement into farrowing. Cleaning typically involves washing with warm water (3—4 L/sow) and disinfection with a mild germicidal solution (Fig. 3.16).

In a hoop barn gestation unit, feeding stalls can be used as wash stations; however, if that is the case, it is desirable that designated stalls with a 1.5% slope toward the front be used to prevent drainage of all the wash water into the bedding pack.

Properly constructed working chutes and ramps (Fig. 3.18) facilitate handling of animals for weighing, sorting, loading, and administration of veterinary treatments (Bishop et. al. 1987). A good working chute design minimizes stress both for the

FIGURE 3.16 Cleaning sows by pressure washing.

Credit: Photo: Loza, A., 2004. CJSC Agro-Soyuz Ukraine - hoop barns in the Ukraine, In: Hoop Barns & Bedded Systems for Livestock Production Conference, 14 September 2004, Ames, Iowa, Agro-Soyuz, the Ukraine. With permission of Honeyman, ISU Hoop Conference.

FIGURE 3.17 Swine sorting and handling in a hoop structure.

Handling of pigs for research purposes—weighing and P2 back fat measurements
(upper). Handling animals in feeding alley for weighing *(middle left)* restraining a pig in
the box for ultrasonic back fat measurements *(middle)* and moving from the sorting pen to
the measurement area with the aide of gates and a hog panel (red) *(middle right)*.
Ultrasound console near the restraining box with an image of the longissimus
muscle at the P2 location *(lower)*.

Credit: Photo by Ioan Hutu, 2016, Horia Cernescu Research Unit, Timisoara, Romania.

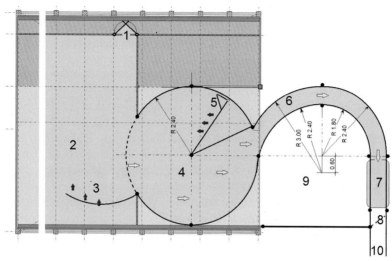

FIGURE 3.18 Working chute located at the end of a hoop structure.

1. Two-way inlet gates; 2. holding pen; 3. moveable entry gate of the round crowding pen and direction of gate movement (red arrows); 4. round pen with direction of animal movement (yellow arrows); 5. 2.40-m-long swinging gate; 6. curved lane or working alley; 7. scales; 8. 90-cm-wide sorting gate; 9. pen for sorting off lightweight animals; 10. to the loading ramp; 20% max slope with 25 cm wide steps.

Credit: Draw by Ioan Hutu, 2017, Swine Experimental Unit, Timisoara, Romania.

animals and for the stock handlers. Reduction of stress on the animals is important to prevent conditions that manifest themselves as subsequent pork quality problems, such as pale, soft, and exudative or dark, firm, and dry product.

Two stock handlers can typically accomplish most animal handling tasks with a well-designed chute system. One handler moves animals into the round forcing pen, and the second handler performs the management procedure required at the time.

A good working chute design is based on the basic biological and behavioral characteristics of swine. These include pigs' wide sight range and relatively good hearing as well as their natural curiosity, which, however, is coupled with a moderate fear of people and of novelty. Swine have panoramic vision spanning 300–310° and most likely are seen in color (Prince, 1977). Their sight discrimination is relatively short range, and for this reason, unfamiliar, dark, or extremely bright areas will deter their movement. The pig's sight range constitutes its personal space. Stockpersons can take advantage of this fact. When a stockperson enters the pig's sight range, it will try to escape.

The direction of escape depends on where within the sight area the stockperson enters—in front of the shoulders will typically result in the pig turning and reversing direction; behind the shoulders will typically result in forward movement. Movement of the stockperson outside the pig's sight range will cause the animal to stop its movement (Grandin, 1989; Morrison and Lee, 2003).

Two other basic swine behavior patterns may be used to advantage in handling swine. These are the animals' propensity for moving in groups and their tendency to follow the leader or other animals in front of themselves (Stolba and Wood-Gush, 1989; Jensen, 1982). A good handling system allows for these behaviors to be expressed by letting animals see, hear, and touch each other so that they can move as a group and follow lead animals.

The handling system can be located at one end of a hoop structure. Animals typically enter the system from a holding pen or the access road at the end of the hoop structure (which is often fenced and can serve as holding pen). Entry is into a round pen (sometimes referred to as a forcing pen), which in turn channels animals into a narrow chute area, which leads to scales, sorting gates, or loading ramps.

The round pen should have a concrete floor with the same color and finish as the concrete floors in the remainder of the hoop barn. Walls of the round pen should be about 110 cm tall and can be constructed of wood, plastic panels, or galvanized steel sheets. The pen typically ranges from a half circle to a three-quarter circle in its extent. A swinging gate equal to the circle's radius is hung from a post at the center point. Stockpersons guide swine into the round pen and then swing the gate behind the group, which forces them to move toward the outlet and into the working chute or alley.

The working alley can be straight or curved (Fig. 3.18). The alley walls should be solid so that animals are not distracted by outside movements and can instead focus on following the animals ahead of themselves. Alley walls can be constructed from the same materials and in the same manner as the round pen walls. The alley width depends on the size of swine being handled at the facility with about 60 cm being desirable for adult animals. Alley walls must be free of any protrusions that might injure animals, and all corners should be gradual and rounded. At the end of the alley, a scale may be installed followed by a sorting gate (90 cm wide). One of the sorting gate outlets may lead to a loading ramp. This ramp should allow only single file movement of animals. Typically ramps have a wooden floor with cleats to provide good traction to minimize slippage and injuries. Alternatively the floor can be a series of steps 25—30 cm wide and 9—10 cm high. The slope of the ramp should be no greater than 20°.

All wall materials and panels in the entire handling area should be of the same color to avoid contrasts that may deter animal movement. In addition, the lighting of the area should be such that animals are moving toward lighter areas (but without glare) as this will facilitate their movement. It is also advisable to avoid excessive noise (Hălmăgean 1984).

3.7 Barn sanitation

Barn sanitation is an important part of disease prevention. As soon as a lot of swine are moved out of the structure, cleanup in preparation for the next group begins. Good sanitation requires control of microbes, insects, and rodents.

3.7.1 Sanitation of barn structure

Cleaning the hoop structure itself is necessary to minimize the presence of pathogens. Cleaning typically involves three steps: (1) removal of manure and other residues (Fig. 3.19); (2) washing (Fig. 3.20); and (3) disinfection.

After animals are removed, electrical power may be disconnected and physical removal of manure and other residues (such as left-over feed) commenced. It may

FIGURE 3.19 Removal of manure from hoop barns.

Swine Experimental Unit—removing the manure with a front-end loader (left) and manually with forks (right).

Credit: Photo by Ioan Hutu, 2019, Horia Cernescu Research Unit, Timisoara, Romania.

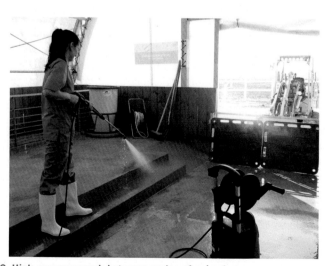

FIGURE 3.20 High pressure wash between cycles of animals.

Swine Experimental Unit—high pressure washing of barn and equipment.

Credit: Photo by Ioan Hutu, 2019, Horia Cernescu Research Unit, Timisoara, Romania.

be useful to wet the manure before removal if it has a dry dusty surface. It may also be useful to spray the manure pack and the building structure with insecticide before manure removal if fly populations are extremely high. Immediately following manure removal is an opportune time for repairs to gates, feeders, etc.

Shortly after manure removal, washing should commence (Fig. 3.19). The purpose of washing is to fully remove any remaining manure residue that adheres to the surfaces of the structure or equipment and therefore effectively remove the majority of pathogenic microorganisms (up to 90%−95%). Microbe removal is accomplished by a combination of the mechanical pressure of the wash stream along with the action of added disinfecting agents. Washing may include two steps:

- an initial low-pressure wash to soak the residue and remove that, which is only lightly adhered;
- a high-pressure wash with detergent to remove all adhered residue.

Hot (60−80°C) detergent solutions may be used if the pressure washing equipment is designed to handle such. Detergents may include ingredients such as calcium carbonate, sodium metasilica, anionic or cationic detergents, or phosphates. Washing should begin at the highest levels of the structure and work downward. A good pressure washer will generate a triangular stream a few millimeter thickness with a nozzle pressure of 10−20 bar. A rinsing step to remove detergent solution may follow the high-pressure wash.

Once washing is completed, disinfection can begin. Disinfection involves the spraying of a germicidal solution onto the surfaces of the structure. This may be initiated as soon as the water from the washing has run off but surfaces are not yet dried. The moist surfaces will typically enhance the effect of the disinfectant. Possible choices of disinfectant may include *Virkon S* applied at 300 ml/m^2 at a pressure of 35 bar with a 45° triangular jet or some older formulas such as 10%−20% slaked lime or 2% formaldehyde.

It is important to exercise constant vigilance for preventing disease introduction into the swine housing structures. Some practices useful for accomplishing that are used of protective clothing and foot baths for workers and strict adherence to all-in-all-out management techniques.

Occasionally, it may become necessary to disinfect/decontaminate a barn, while swine are present to prevent respiratory disease (caused by organisms such as *Pasteurella* spp., *Klebsiella* spp., *Mycoplasma* spp., or *Heamophilus* spp.). In such instances, a 0.5%−1.0% solution of a suitable disinfectant may be misted into the structure at a rate of 1 L/100 m^3 twice per day. Droplet size of such a mist should be 50−150 μm. Ventilation rates should be increased during such procedures. In fact, oftentimes, increasing ventilation alone may prevent the respiratory outbreaks. In addition, exposure to ultraviolet light or aerosols of negatively charged ions may be beneficial.

3.7.2 Insect control in deep bedded structures

Insect control targets flies, mosquitoes, ticks, and other potential disease-carrying species. To limit insect populations, the following practices should be observed:

- manure storage areas located a minimum of 100 m from barns;
- water and waste runoff pits be periodically emptied;
- dead animal composting sites be located away from barns;
- any standing water areas in the vicinity of the farm be eliminated;
- bedding moisture be maintained below 75%; and
- no garbage disposal into barns.

Successful insect control depends on adhering to the above recommendations as well as effective use of insecticides of proper composition and concentration.

An initial insect control step may be made immediately upon removal of the group of swine by applying an insecticide (e.g., Divipan as an aerosol at a rate of $2-4$ g/m^3). Contact time of such a treatment should be 24 h after which manure removal, washing, and disinfecting can be done. It may be necessary to test for effectiveness of insecticides if resistance has developed to certain products. Typically, 1% Alfacron, 3% Neguvon, or 1% Safrotin is effective, however. A second insect control step may be taken following disinfection of the barn using a long residual insecticide such as 3% Clorofos or 2% Alfacron. If barns are sprayed for insects while swine are present, insecticides such as 1% Alfacron should be used about once every 3 weeks (Decun, 1995). It is also important that ventilation be increased during application of insecticides while swine are present, to prevent accidental poisoning of the animals.

3.7.3 Rodent control

Rodent control targets rats, mice, and other rodents to prevent the damage they cause and the spread of diseases they may carry. There are a number of approaches to rodent control. These include mechanical (such as trapping or flooding of their burrows), chemical (such as inorganic, organic, or synthetic rodenticides), or biological (such as use of wild or domestic predators). Effective rodent control includes the following aspects:

1. Prevention of population buildup by limiting access to buildings with appropriate barriers and by eliminating potential attractants such as waste feed buildup, etc.
2. Use of effective rodenticides (baits). In order to prevent development of resistance to baits, it is advisable to use a variety of rodenticides with varying mechanisms of action. It is more effective to rotate baits than to increase dosage of a given bait if resistance begins to be a problem. A typical baiting procedure would be to begin with a slow action baits such as anticoagulants such as furfuril hydramide or derivatives of coumarin. After $4-6$ days, these prebaits are removed and replaced with very toxic fast-acting baits such as hydrogen

phosphide gas, zinc phosphate, or aluminum phosphate. If the baits are not strong enough, the rodents may develop an alarm reaction and vacate the building only to relocate in another nearby area. Bait distribution using meaty substances (fish, chicken, beef, pork, etc.) is very effective as are liquid baits placed in water.

3. Rodent control requires a concerted and persistent effort. Baits must never be allowed to deplete during the control effort. Control efforts should be undertaken over the entire farmstead or even multiple nearby farmsteads simultaneously to prevent rodents from merely migrating from one site to another.

Once rodent control has been achieved, all baits should be removed and destroyed and all burrows and tunnels and other access points effectively stopped up. Periodic inspection for reinfestation by rodents should be performed. If reinfestation does occur, a new round of control measures should be introduced using different baits and procedures than used in the initial effort.

3.8 Manure removal and disposal

The frequency of manure removal from deep bedded hoop barns depends on the type of animals maintained in the structure and the amount of bedding used. Typically for gestation units, manure will be removed from one to four times per year. For grow-finish units, it is common practice to remove manure after each lot of hogs is marketed (approximately twice per year). For climatic conditions similar to Romania's, the quantity of manure from a 180−200 head grow-finish unit will also vary depending on season. During winter, more bedding is used and manure amounts will be approximately 640 kg per sow (Santonja et al., 2017), 600 kg per hog, or 830 m^3 for the entire structure at the end of each finishing cycle. During summer, typical amounts would be 300 kg per hog or 415 m^3 per barn (Fulhage, 2003).

Skid-steer loaders fitted with a grapple fork and bucket are perhaps the best implement for manure removal although any tractor with a front-end loader will work. Cleaning will take from 12 to 15 h per hoop barn depending on the equipment used and amount of manure present (Honeyman, 1998; Honeyman et al., 1999). It is desirable during winter to delay cleaning until just before introduction of the next lot of piglets as the manure pack will help maintain the barn at a higher temperature during its down time.

Manure can be spread on cropland immediately following removal or stored for later spreading depending on field work schedules. Direct spreading is common practice. In wean-to-finish units where the growth period requires about 6 months, the manure pack is already significantly composted as a result of microbial action in the pack during this time (Tiquia et al., 2000; Fulhage, 2003). One ton of manure will contain 7−16 kg N, 4.5−9 kg P_2O_5, and 5.5−10.5 kg K_2O. A common application rate (Fig. 3.21) would be 5−8 tons per *ha* per year or 15−25 tons per *ha* applied once every 3 years (Kruger et al., 2006).

FIGURE 3.21 Loading and spreading composted manure on cropland.

Loading composted manure into a manure spreader with a front-end loader (left).
Spreading compost on the field (right).

Credit: Photo by Ioan Hutu, 2019, Banat University of Agricultural and Veterinary Medicine Farm, Timisoara,
Romania.

Composting results from microbial degradation of the organic material in the
manure pack. There are a number of composting methods, but aerobic composting
requires a minimum of labor and investment (Rynk, 1992). Manure may be placed in
long piles (windrows) 1.2—2.0 m high and 3.0—4.0 m wide (Richard et al., 1998).
For the composting process to proceed effectively, the proper *C:N* ratio is required.
If carbon is too high, the composting process will not proceed to completion. If nitrogen is too high, excess ammonia will be produced. Moisture content of the
manure should be between 50% and 60% for optimum composting. Windrows of
manure should be aerated by turning or mixing periodically to maintain aerobic conditions and the corresponding optimum temperature of about 55°C within the pile
(Paul, 2005). Microbial composting activity will cease if moisture falls below
40% or temperature below 30°. Composting produces a product with a *C:N* ratio
of *12:1*, and the volume of the product will drop to about 40% after 6 weeks (Tiquia
et al., 2000). As a consequence, storage space requirements are relatively low; however, during rainy weather, some leaching[3] of *N* may occur from compost stored on
the ground or runoff may occur from compost stored on concrete pads (Richard
et al., 1998). Depending on season and rainfall amounts, composting will result in
variable losses of plant nutrients. 3—60% of *N* and 20%—45% of *P* and *K* may be
lost (Garrison et al., 2001). Composting also renders most weed seeds sterile
(Eghball and Lesoing, 2000) and makes land spreading easier as a result of the
decreased volume. Compost breaks up better than fresh pack manure and thus
spreads more evenly on the land as well. As indicated by Table 3.3, both fresh
manure and compost are excellent sources of important soil nutrients and organic

---

[3] Penetration of *N* from the compost into the soil and potentially into the underground water aquifer—
more likely during periods of heavy rainfall as *N* compounds are readily water soluble.

Table 3.3 Comparative composition of corn fresh manure and compost[a]

Type	Ash g/kg	Total P g/kg	Total K g/kg	Total C g/kg	Total N g/kg	C/N	NH$_4$-N µg/g	NO$_3$-N µg/g
Manure composition over fall season								
Fresh	345	11.4	21.1	323	26.7	12.2	2205	17
Compost	616	9.6	16.8	190	17.1	11.1	430	785
Manure composition over spring season								
Fresh	360	8.1	18.9	330	26.2	12.8	2165	87
Compost	661	7.8	16.0	175	14.6	11.9	835	445

[a] All values expressed on a dry matter basis except for water content.
Adaptation after Loecke, T.D., Liebman, M., Cambardella, CA., Richard, T.L., 2004. Corn response to composting and time of application of solid swine manure, Published in Agron. J. 96:214–223.

matter (Loecke et al. 2004). Mismanagement of this resource can, however, lead to pollution of both air and water (Sharpley et al. 1998; Zebarth et al. 1999). Another potential strategy for handling swine manure is to use it as feedstock for biogas generation (methane). Each animal from weaning to market will produce 720 m^3 of biogas or the equivalent of 2 kWh (12.6 MJ or 3009 kcal). Typically a production unit with 1000 head grow-finish capacity is the minimum size required to justify investment in biogas generation facilities (Sexton, 2002).

References

Bishop, G.E., Turnbull, J.E., Kains, F.A., Grandin, T., 1987. Site and Building Planning for Swine Production Canada Plan Service. NEW 87:06, Plan − 3002.

Brumm, M.C., Harmon, J.D., Honeyman, M.S., Kliebenstein, J.B., Zulovich, J.M., 1999. Hoop Structures for Gestating Swine. AED44. MidWest Plan Service, Iowa State Univ., Ames, IA, 16pp.

Brumm, M.C., Dahlquist, J.M., Heemstra, J.M., 2000. Impact of feeders and drinker devices on pig performance, water use, and manure volume. Swine Health Prod. 8, 51−57.

Carr J. Garth Pig Stockmanship Standards. 5M Enterprises. 1998 https://www.amazon.com/Garth-Stockmanship-Standards-John-Carr/dp/0953015092.

Decun, M., 1995. In: Sanitatie Veterinară. Helicon, Timişoara.

Dinu, I., 1978. Influenţa meiului asupra producţiei la porcine, (Ed.), Ceres, Bucuresti.

Eghball, B., Lesoing, G.W., 2000. Viability of weed seeds following manure windrow composting. Compost Sci. Util. 8, 46−53.

Fulhage, C., 2003. Manure management in hoop structures nutrients and bacterial waste. Nutrients and Bacterial Waste, MU Guide, EQ 352. Published by MU Extension, University of Missouri-Columbia, muextension.missouri.edu/xplor/.

Garrison, M.V., Richard, T.L., Tiquia, S.M., Honeyman, M.S., 2001. Nutrient losses from unlined bedded swine hoop structures and associated window composting site. In: ASAE Meeting Paper 01−2238. ASAE, St. Joseph, MI.

Grandin, T., 1989. Behavioral principles of livestock handling. Prof. Anim. Sci. 5 (2), 1–11.

Hălmăgean, P., 1984. Tehnologia creşterii şi exploatării porcinelor, (Ed.), Ceres, Bucureşti.

Halverson, M., Honeyman, M.S., Adams, M., 1997. Swedish Deep-Bedded Group Nursing Systems for Feeder Pig Production. AS-12. Iowa State University Extension, Ames, IA.

Harmon, J.D., Xin, H., 1997. Thermal Performance of a Hoop Structure for Finishing Swine. ASL-R1391. 1996 Swine Research Reports. Iowa State University Extension, Ames, IA, pp. 104–106.

Harmon, J.D., Honeyman, M.S., Kliebenstein, J.B., Richard, T., Zulovich, J.M., 2004. Hoop Barns for Gestating Swine. AED44, Rev. Ed. MidWest Plan Service. Iowa State Univ., Ames, IA, p. 20.

Honeyman, M.S., 1997. Hooped Structures with Deep Bedding for Grow-Finish Pigs. ASL-R1392. 1996 Swine Research Reports. Iowa State University Extension, Ames, IA, pp. 107–110.

Honeyman, M.S., 1998. Alternative Swine Housing: A Brief Review of Tarp-Covered Hooped Structures With Deep Bedding for Grow-Finish Pigs. December, 2pp. www.depts.ttu.edu/porkindustryinstitute/articles/alternat.htm.

Honeyman, M.S., Harmon, J.D., 2003. Performance of finishing pigs in hoop structures and confinement during winter and summer. J. Anim. Sci. 81, 1663–1670.

Honeyman, M.S., Kent, D., 2000. Reproductive Performance of Young Sows from Various Gestation Housing Systems, 1999 Swine Research Reports, AS-642. Iowa State University, Ames, IA.

Honeyman, M.S., Koenig, F.W., Harmon, J.D., Lay Jr., D.C., Kliebenstein, J.B., Richard, T.L., Brumm, M.C., 1999. Managing Market Pigs in Hoop Structures. Pork Industry Handbook PIH-138. Purdue Univ., W. Lafayette, IN.

Honeyman, M.S., Harmon, J.D., Kliebenstein, J.B., Richard, T.L., 2001. Feasibility of hoop structures for market swine in Iowa: pig performance, pig environment, and budget analysis. Appl. Eng. Agric. 17 (6), 869–874.

Honeyman, M.S., Harmon, J.D., Kent, D., Hummel, D., Christian, L., 2002. The Effects of Gestation Housing on the Reproductive Performance of Gestating Sows. A Progress Report ISRF01-12. Iowa State University, Armstrong Research and Demonstration Farm.

Houghton, D., 1997. ArchDeluxe — research may fine-tune feeding in hoop finishers. Hogs Today 2. January.

www.ae.iastate.edu.

Jacobson, L.D., Hetchler, B.P., David, R., Schmidt, Johnson, V.J., 2004. Ammonia, hydrogen sulfide, odor, and pm10 emissions from deep-bedded hoop and curtain-sided pig finishing barns in Minnesota, alternative swine housing systems. In: An International Scientific Symposium, September 15, 2004 Ensminger Room. Kildee Hall, Iowa State University Ames, Iowa.

Jannacsh, R., 1996–7. Happy Hogs, Sustainable Farming. winter, pp. 2–7.

Jensen, P., 1982. An analysis of agonistic interaction patterns in group-housed dry-sows-aggressive regulation through an avoidance order. Appl. Anim. Ethol. 9, 47.

Kruger, I., Taylor, G., Roese, G., Payne, H., 2006. Deep-litter Housing for Pigs. In: PRIME-FACT, vol. 68. NSW Department of Primary Industries, State of New South Wales. http://www.dpi.nsw.gov.au.

Loecke, T.D., Liebman, M., Cambardella, C.A., Richard, T.L., 2004. Corn response to composting and time of application of solid swine manure. Published in Agron. J. 96, 214–223.

Loza, A., 2004. CJSC Agro-Soyuz Ukraine — hoop barns in the Ukraine. In: Hoop Barns & Bedded Systems for Livestock Production Conference, 14 September 2004, Ames, Iowa.

Maltman, J., 2004. Manitoba agriculture, food and rural initiative, hoop barns in Canada. In: Hoop Barns & Bedded Systems for Livestock Production Conference, 14 September 2004, Ames, Iowa.

Messenger, J., 1996. Alternative structures make their mark. Pork 96, 20–23.

Metcalfe, J.P., 1999. Guidance on the Control of Noise on Pig Units. ADAS, Defra.

Morrison, R., Lee, J., 2003. Handling and sorting pigs in large groups housed in deep-litter systems. In: Second International Symposium on Swine Housing Held in Research Triangle Park, NC on October 12–15.

Onan, G.W., Mircu, C., Irina, P., Hutu, I., 2016. Hoop structure for wean to finish pigs: management and gender influence in a summer trial. Lucrări ştiinţifice - Medicină veterinară, Editura Ion Ionescu de la Brad Iaşi 59 (3), 388–394.

Olsson, A., Svendsen, J., Andersson, M., Rantzer, D., Lenkens, P., 1992. Group housing systems for sows, I - electronic dry sow feeding on Swedish farms, an evaluation of the use of the system in practice. Swed. J. Agric. Res. 22, 153.

Paul, J.W., 2005. Composting hog manure - why isn't everyone doing it? Adv. Pork Prod. 16, 263–268.

Payne, H., 2004. Deep-litter housing systems an Australian perspective. In: Hoop Barns & Bedded Systems for Livestock Production Conference, 14 September 2004, Ames, Iowa.

Prince, J.H., 1977. The eye and vision. In: Swensen, M.J. (Ed.), Dukes Physiology of Domestic Animals. Cornell University Press, New York.

Richard, T.L., Harmon, J., Honeyman, M., Creswell, J., 1998. Hoop Structure Bedding Use, Labor, Bedding Pack Temperature, Manure Nutrient Content, and Nitrogen Leaching Potential. ASLR1499, In: 1997 ISU Swine Research Report, AS-638. Dept, of Animal Science, Iowa State University, Ames, IA, pp. 71–75.

Rossiter, L.T., Honeyman, M.S., 1997. Hoop Farrow-to-finish Demonstration Iowa State University, Northwest Research Farm and Allee Demonstration Farm. ISRF97-29.31. http://www3.abe.iastate.edu/hoop_ structures/swine/archives/other/systems.htm.

Rynk, R. (Ed.), 1992. On-farm Composting Handbook. Northeast Regional Agric, Eng, Serv., Ithaca, NY.

Santonja, G.G., Georgitzikis, K., Scalet, B.M., Montobbio, P., Roudier, S., Sancho, L.D., 2017. Best Available Techniques (BAT) Reference Document for the Intensive Rearing of Poultry or Pigs. EUR 28674 EN. Publications Office of the European Union, Luxembourg.

Sexton, B., 2002. Biogas — an Overview, Alberta Agriculture, Food and Rural Development. http://www.agriCgov.ab.ca/engineer/biogas.pdf.

Sharpley, A., Meisinger, J.J., Breeuwsma, A., Sims, J.,T., Daniel, T.C., Schepers, J.S., 1998. Impacts of animal manure management on ground and surface water quality. In: Hatfield, J.L., Stewart, B.A. (Eds.), Animal Waste Utilization: Effective Use of Manure as a Soil Resource. Ann Arbor Press, Ann Arbor, MI, pp. 173–242.

Stolba, A., Wood-Gush, D.G.M., 1989. The behaviour of pigs in a semi-natural environment. Anim. Prod. 48, 419–425.

Stuhec, I., Kovac, M., Malovrh, S., 2002. Efficient heating of piglet nests. Archiv für Tierzucht 45 (5), 491–499.

Tiquia, S.M., Richard, T.L., Honeyman, M.S., 2000. Effect of windrow turning and seasonal temperatures on composting of hog manure from hoop structures. Environ. Technol. 20 (9), 1037–1046.

Voyles, R., Honeyman, M.S., 2006. Absorbency of Alternative Livestock Bedding Sources. ASL-R2153, Animal Industry Report AS-652. ISU Ext. Serv., Ames, IA.

Zebarth, B.J., Paul, J.W., Van Kleeck, R., 1999. The effect of nitrogen management in agricultural production on water and air quality: evaluation on a regional scale. Agric. Ecosyst. Environ. 72, 35–52.

Animal management and welfare

4.1 Temperature

During the early stages of life, piglets require an environment with a warm and well-regulated temperature range as they have not yet developed their thermoregulatory mechanisms very well. As they mature, these mechanisms begin to function more effectively.

Thermoregulation allows animals to control their internal body temperature in environments that are outside their thermoneutral range. Thermoregulation may involve thermogencsis (i.e., heat production to keep body temperature at normal levels in cold environments) or thermolysis (i.e., loss of body heat to the environment when temperatures are high). These mechanisms allow the animal to maintain its core body temperature near the physiological ideal which for swine is 38.6−39.2°C.

4.1.1 Critical temperature and heat production

As indicated in Fig. 4.1, life is possible within an ambient temperature range bounded by Tdc (minimum low temperature) and Tdw (maximum high temperature). The thermoregulatory zone (or homeothermal zone), bounded by Thi and Ths, is a smaller region within that range.

Within the thermoregulatory zone, the animal is able to maintain its core body temperature at a physiological level. Within that range, in turn, is a smaller region called the thermoneutral zone bounded by Tci and Tcs, the lower and upper critical temperatures, respectively.

Within this zone, the animal can maintain its core body temperature at a physiological level without any thermoregulation—minimal expenditure of energy for heat loss or heat production (Verstegen and Close, 1994).

Critical temperature is dependent on breed, feeding level, group size, and other animal welfare and environmental factors (Close 1981). Lower critical temperature rises as group size decreases, feeding levels drop, ventilation speed increases, bedding decreases, or body surface moisture increases. Upper critical temperature rises with increases in ventilation speed and with the addition of body surface moisture such as by misting animals with a water spray (Carr, 1998). When ambient

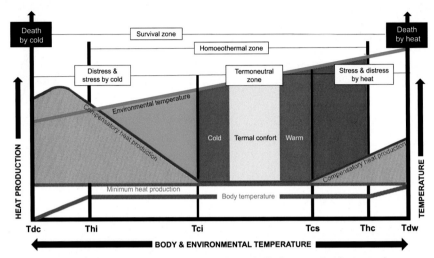

FIGURE 4.1 Regulation of body temperature and metabolic heat production at various environmental temperatures.

Body temperature—*red line*. Metabolic heat production (minimum and compensatory)—*mauve lines*. *Tdc*—temperature at which death by hypothermia occurs; *Thi*—homoeothermic lower temperature (lower extreme of thermoregulatory zone)—increasingly lower environmental temperature will cause a decrease in body temperature; *Tcs*—low critical temperature (lower end of thermoneutral zone); *Tci*—high critical temperature (upper end of thermoneutral zone); *Ths*—homoeothermic high temperature (upper extreme of thermoregulatory zone)—increasingly higher environmental temperature will cause increasing body temperature; *Tdw*—temperature at which death by hyperthermia occurs.

Redrawn after Dinu, I., Influenta mediului asupra productiei la porcine, Ceres (Ed.) Bucuresti, 1978 and Yousef, M. K., 1985. Thermoneutral zone. In: Yousef, M. K. (Ed.), Stress Physiology in Livestock, vol. I., CRC Press, Boca Raton, FL. pp. 47–54.

temperature is below the lower critical temperature, the animal's body initiates thermogenic processes to add heat to the body core.

When ambient temperature is above the upper critical temperature, the animal's body initiates mechanisms that disperse heat from the body.

Ambient temperatures outside the thermoneutral zone of the animal will lead to losses of performance because of the energy diverted to temperature control (Fig. 4.1). The body mechanisms that result in heat losses include evaporation due to respiration and sweating (the latter is not effective in swine), losses in elimination products (fecal and urinary), and losses due to radiation, convection, and conduction (Burlacu et al. 2005).

Mechanisms that allow for thermogenesis within the animal's body include an increased basal metabolic rate, heat from digestion of feed, and increased muscle

Table 4.1 Distribution of heat loss from swine weighing 25–220 kg.

Air temperature (°C)	Percentage of total loss (%)			
	Radiation	Convection	Conduction	Vaporization
3–5	32%–38%	34%–40%	11%–15%	13%–17%
20–22	24%–30%	31%–37%	9%–13%	25%–31%
37–39	2%–4%	4%–6%	2%–4%	85%–93%

Adaptation after Burlacu, R., Grossu, D. V., Marinescu, A. G., Burlacu, G. M., 2005. Modelarea matematica a proceselor de metabolism energetic si proteic la suine, INBA, Balotesti, Ed. Cartea Universitara, Bucuresti.

activity both voluntary and shivering. Table 4.1 illustrates the relative contributions of each of these various processes to heat loss from the body at different ambient temperatures.

4.1.2 Factors affecting temperature neutrality

Temperature neutrality depends on body mass (more specifically the ratio of surface area to volume, which increases at lower weights), type of housing (group size, presence of bedding, etc.), plane of nutrition (feeding level), air current speed, and relative humidity. Plane of nutrition directly affects body heat production and therefore has profound effect on the thermal neutral range. Table 4.2 indicates the interaction of plane of nutrition and body mass on thermal neutral range for swine. Group housing will also affect thermal neutral range (Holmes and Close, 1977).

As indicated in Table 4.3, animals housed in groups require less feed energy intake to meet maintenance requirements. Table 4.3 also indicates in the right-hand column the amount of extra feed intake required for each degree C below

Table 4.2 Thermoneutral zone (°C) for single or group animals as a function of feeding level (EM).

Class of pig or weight	Feeding level		
	1x ME[a]	2x ME	3x ME
Piglets	32±1°C	31±2°C	29±2°C
Weanlings	29±3°C	26±5°C	23±6°C
Growing hogs	28±4°C	25±5°C	22±6°C
Finishing hogs	27±4°C	24±5°C	20±6°C
Lean gestating sows	25±5°C	21±6°C	18±7°C
Fat gestating sows	24±6°C	20±7°C	16±8°C

[a] *ME is maintenance metabolizable energy = 100 kcal/kg$^{0.75}$/day.*
Adaptation after Holmes, C. W., Close, W. H., 1977. The influence of climatic variables on energy metabolism and associated aspects of productivity in pigs. In: Haresign, W., Swan, H., Lewis, D., Nutrition and the Climatic Environment. Butterworths, London, pp. 51–74 and Dinu, I., 1978. Influenta mediului asupra productiei la porcine, Ceres (Ed.), Bucuresti.

Table 4.3 Extra feed intake necessary to meet maintenance requirements depending on housing for each °C below the lower critical temperature.

Class of pig or weight	Type of housing or body condition	Extra feed[a] (g/°C/day)	Needed EM (kcal/°C/day)
Piglets	Individual	3.68	11.60
Weanlings	Individual	12.89	40.60
	Group	11.97	37.70
Growing hogs	Individual	23.02	72.50
	Group	22.10	69.60
Finishing hogs	Individual	33.14	104.40
	Group	32.22	101.50
Gestating sows	Lean	54.32	171.10
	Fat	31.30	98.60

[a] *Data adapted after Holmes, C. W., Close, W. H., 1977. The influence of climatic variables on energy metabolism and associated aspects of productivity in pigs. In: Haresign, W., Swan, H., Lewis, D., Nutrition and the Climatic Environment. Butterworths, London, pp. 51–74 and Luca, I., 2000. Curs de alimentatia animalelor, Marineasa (Ed.), Timişoara, for feed with metabolizable energy density of 3150 kcal/kg.*

the lower critical temperature (Holmes and Close, 1977). Stage of production (animal weight and age) and housing floor surface also have profound effects on the thermal neutral zone. Neonatal piglets are the most sensitive to temperature as their thermoregulatory mechanisms are not yet well developed.

The thermal neutral range for newborns is 35–41°C, whereas by the time they attain a 2 kg weight, the range drops to 29–35°C (Draghici 1991). For 5 kg weaned piglets housed on straw bedding, the thermal neutral zone is 27–30°C; at 10 kg weight, it is 20–24°C; and at 20 kg weight, it is 15–20°C. Group housing on straw bedding decreases the lower critical temperature by 1–3°C.

For adult breeding swine, there is some variance in thermal neutral zone depending of stage of production. Gestating sows have a range of 10–29°C (Holden et al. 1996). For nursing sows, the thermal neutral range is narrower, 16–24°C (Brent 1986), or perhaps even as narrow as 18+1°C (Carr 1998).

As air movement increases, heat loss from animal bodies increases due to convection. Young swine are most sensitive to this effect. The heat loss from convection is dependent on animal weight, group size, ambient temperature, and duration of air movement (Sallvik; Walberg 1984). In cold barns, 60 kg group-housed hogs experiencing air speeds of 0.21–0.30 m/s will have a lower critical temperature 1°C higher than in still air. If housed in individual stalls, increases of only 0.05 m/s in air speed will increase the lower critical temperature by the same amount (Verstegen and van der Hel 1976; Close 1981, 1987).

In warm housing, increases in air speed will increase the upper critical temperature. Higher relative humidity inhibits an animal's ability to disperse excess heat through evaporative loss. For example, at 30°C, swine body temperature will rise by 1°C for each incremental increase of 18% relative humidity (Holmes; Close 1977).

In spite of all the data and recommendations, the ultimate test of thermal comfort is the behavior of the animals.

In cold barns and low ambient temperatures, swine will group together to reduce total surface area and therefore reduce radiative, convective, and conductive heat losses to the environment. Increasing feeding level will help animals generate the added metabolic heat to maintain their thermoregulatory zone (see Fig. 4.1 and Table 4.2).

In warm barns and high ambient temperatures, respiration rate will increase significantly to enhance evaporative heat losses. Sows may reach respiration rates of 120 cycles per minute to maintain body temperature (Draghici 1991). Increasing airflow rates will facilitate removal of heat from animal bodies by increased convective skin losses and increased evaporative losses from respiratory tissues, resulting in a greater comfort level for the animal.

4.2 **Nutrition**

Nutrition of swine reared in hoop barns is dependent on a variety of factors.

4.2.1 **Feed nutrients and energy**

Optimal nutrition involves balancing of feed ingredients and nutrients to meet the requirements of the animal for basic maintenance of physiological function, for growth, and for reproduction/lactation. The critical nutrient groups include energy components, amino acids, vitamins, and minerals. Energy components (carbohydrates and lipids) are most important as each organ or organ system requires energy to perform its function. If energy intake is not adequate, the body will sacrifice growth and reproductive functions to provide for basic maintenance (Crainiceanu and Matiuti, 2006; Burlacu et al., 2002). If energy intake is more than needed for body functions, excess calories will be stored as fat.

Energy content of swine feeds is generally expressed as digestible energy (DE) or metabolizable energy (ME) in either kcal or MJ[1] (Kopinski et al. 2005). Typically, the energy requirement for maintenance is about 191−216 kcal ME/$kg^{0.60}$ of body weight for growing and finishing pigs and about 100−110 kcal ME/$kg^{0.75}$ for gestating and lactating sows (NRC, 2012). Growth, lactation, gestation, and thermogenesis require extra energy over and above this amount.

Amino acids are another important nutrient and are the basic constituents of protein. There are 20 amino acids found in animal tissues, swine require 10 of these (the essential amino acids—ones that the body cannot synthesize) be present in the diet.

---

[1] 1000 Kcal = 4.1868 MJ.

For cereal grain-based diets, lysine, threonine, tryptophan, methionine, and isoleucine are typically deficient.

They must therefore be supplemented by feeding excess total protein or by supplementation with synthetic forms to achieve the proper ratios among the essential amino acids. If the ratios are out of balance or if even one essential amino acid is slightly deficient, the result will be akin to a deficiency of all of them. Table 4.4 indicates the amount of lysine required for various swine production stages and the ideal ratios of the remaining essential amino acids to lysine.

Vitamins serve a metabolic regulators within the animal's body and are therefore critical for normal health, growth, reproduction, and lactation. Of particular importance are vitamin D for young swine and vitamins D and E for breeding swine (Hafez, 1993).

Minerals, both microminerals and macrominerals, also serve critical functions for controlling metabolism as well as formation of the skeleton. Typically, Na, Ca, and P levels in a ration are balanced individually, while microminerals such as Fe, Zn, Cu, and Mg are supplied with premixed mineral supplements.

Other feed nutrients of special note include cellulose (fiber) and water. Cellulose utilization is dependent on the production stage of the swine. No more than 2% cellulose should be present in diets offered to nursing piglets as they cannot digest it and there is evidence that it prevents mucus formation in the gut, therefore inhibiting one of the gut's important mechanisms for resisting bacterial challenges. 3%–4% cellulose can be present in diets of weaned piglets and the level can rise to 5%–6% for finishing hogs (Luca 2000). Adult breeding swine need a minimum of 4% cellulose, which serves as an efficient agent for reducing hunger in a cost-effective manner.

Water intake for swine is generally about 1.8–3.5 L/kg feed consumed (Brumm et al., 2000, Kruger et al., 2006). Daily water requirements for various stages of production during summertime, or whenever ambient temperatures are above 10–12°C, are given in Table 3.1. As air and water temperatures rise, water consumption also rises as a consequence of the animals making extensive use of evaporative heat losses to keep their core body temperature within physiological limits (Taylor et al. 2006). For example, at 38°C, 89% of heat loss is through evaporative means, particularly from the respiratory tract (Fig. 4.1).

4.2.2 Factors affecting ration formulation

A balanced ration contains optimal quantities and ratios of the various nutrients required by the animals depending on their genetic makeup, stage of production, and environmental factors. Following are some guidelines for ration formulations for swine raised in hoop shelters or other ambient temperature housing facilities.

Energy and protein, the two largest components of the ration, must be in a proper ratio to each other for maximum efficiency of utilization. The recommended amounts and ratios of the essential amino acids to each other and to digestible energy should be like Tables 4.4 and 4.5.

Table 4.4 Lysine requirements and ideal ratio of lysine to the other amino acids in Standardized Ileal Digestible (SID) basis.

Age, category, and weight	Growing and finishing hogs, kg[a]							Gestating sows (days)				Lactating sows[d]
								Small[b]		Large[c]		
	5–7	7–11	11–25	25–50	50–75	75–100	100–135	<90 days	>90 days	<90 days	>90 days	
Lysine g/day[e]	4.0	6.3	11.1	14.8	17.9	18.3	16.9	10.6	16.7	6.7	11.5	48.9
Amino acids ratio to lysine, %												
Arginine, %	45	46	46	46	46	46	46	53	53	52	52	55
Histidine, %	35	35	34	34	35	34	34	35	32	33	30	40
Isoleucine, %	50	51	51	52	53	53	54	58	53	60	51	56
Leucine, %	100	100	100	101	101	101	102	91	93	96	97	113
Methionine, %	30	29	29	29	29	29	29	28	28	27	28	27
Methionine + cysteine, %	55	56	55	56	57	57	59	64	65	72	70	53
Phenylalanine, %	58	59	59	59	60	60	61	55	54	58	57	54
Phenylalanine + tyrosine,%	93	92	93	93	94	95	96	95	95	100	98	112
Threonine, %	58	59	59	60	62	63	66	72	69	84	77	64
Tryptophan, %	18	16	16	17	17	17	13	18	19	21	22	19
Valine, %	63	63	64	65	65	66	67	71	71	78	75	85

[a] Mean whole body protein deposition of 135 g/day between 25 and 125 kg with sex ratio 1:1.
[b] Carrying 12.5 piglets, 140 kg at mating and gaining 65 kg during pregnancy.
[c] Carrying 13.5 piglets, 205 kg at mating and gaining 45 kg during pregnancy.
[d] 210 kg lactating sow feeding average of 11.5 piglets for 21 days gaining an average of 230 g per piglet per day.
[e] For mixed feed based on corn and soy 3400 kcal DE/kg (3265 kcal ME/kg).

Compilation after National Research Council (NRC), 2012. Nutrient Requirements of Swine, National Academies Press, Washington, DC, USA.

Table 4.5 The energy demand for different animal categories at various ambient temperatures (NRC, 2012).

Nutrient	Temperature	Growing pigs, kg[a]							Breeding swine, kg[a]				Gestating sows		Lactating
		5–7[b]	7–11[b]	11–25	25–50	50–75	75–100	100–135	♂♂ 75–100	♀♀ 75–100	♂♂ 100–135	♀♀ 100–135	Small[c]	Large[d]	Sows[e]
Metabolizable energy consumption, kcal/day	Comfort	904	1592	3308	4920	6898	8101	9002	7522	7758	8295	8549	7203	7203	20,731
	+30°C	–	–	2588	3661	4886	5453	5613	5066	5224	5273	5435	7203	7203	14,605
	+25°C	–	–	3123	4456	6075	6950	7404	6455	6652	6955	7170	7203	7203	18,409
	+20°C	–	–	3427	4920	6818	7939	8645	7674	7610	8116	8371	7203	7203	20,068
	+15°C	–	–	3527	5105	7072	8260	9204	7666	7912	8633	8908	7803	8003	20,731
	+10°C	–	–	3788	5527	7881	9372	10,353	8700	8983	9717	10,032	8553	9003	20,731
	+5°C	–	–	4001	5915	8692	10,579	12,029	9822	10,146	11,281	11,640	9303	10,003	20,731
	0°C	–	–	4249	6286	9252	11,281	12,827	10,423	10,836	11,991	12,456	10,381	11,311	–
	–5°C	–	–	4498	6659	9812	11,982	13,624	11,025	11,527	12,700	13,270	11,463	12,619	–
	–10°C	–	–	4746	7031	10,372	12,683	14,421	11,625	12,217	13,410	14,086	12,541	13,926	–
CP, %[f]		30.8	29.1	22.2	21.1	18.5	16.0	13.3	16.7	16.0	15.3	13.6	14.0	11.5	16.3
Lysine, g/day		3.93	6.50	10.30	14.60	17.90	18.20	16.90	18.60	18.40	18.80	17.20	11.20	9.20	49.70
Calcium, g/day		2.26	3.75	6.78	9.77	12.41	13.12	12.76	17.28[h]	16.62	18.02	18.29	12.97	13.45	42.60
Digestible P, g/day		1.19	1.89	3.16	4.55	5.77	6.10	5.94	7.44	7.19	7.95	7.92	5.64	5.85	21.30
FC, g/day[g]		0.288	0.508	1.055	1.569	2.20	2.58	2.87	2.40	2.48	2.65	2.73	2.30	2.30	6.61
ME/lysine ratio		230	244	321	337	385	455	532	404	421	441	497	643	782	417

[a] Mixed gender (of pigs with high-medium lean growth rate (300 g/day of carcass fat-free lean) from 25 to 125 kg body weight).

[b] 2012 NRC Model assumes piglets are in thermoneutral conditions.

[c] Assumes pregnant sows weighing 140 kg at breeding and gaining 66 kg during gestation period; estimated litter size of 13.5 piglets, limit fed and group housed on straw bedding.

[d] Assumes pregnant sows weighing 205 kg at breeding and gaining 50 kg during gestation period; estimated litter size of 13.5 piglets, limit fed and group housed on straw bedding.

[e] Assumes a 180 kg sow losing 4 kg during a 21-day lactation with a litter of 11.5 piglets averaging 230 g daily gain per piglet.

[f] Approximate content of crude protein for a typical corn/soy ration with 3300 kcal ME/kg.

[g] Feed consumption with 3300 kcal ME/kg feed and ambient temperature in the comfort range.

[h] Ca and P recommendations for developing boars and gilts from National Swine Nutrition Guide, USA.

Compilation after NRC and NS Nutrition Guide, USA.

Nutrient availability (or digestibility) is also a concern. In practice, it is not the total amount of a nutrient in the ration that is important, but rather the total amount of digestible nutrient that is present. Of particular concern is unavailability of nutrients due to high levels of fiber or due to excessive heating of ingredients during manufacture or storage. Lysine, for example, can be made relatively undigestible by processing temperatures that are too high. One can ensure adequate availability through using protein from a variety of sources in the ration or by incorporating synthetic amino acids.

Phosphorus is another nutrient that is often not highly available depending on its source. Phosphorus may be supplied via inorganic phosphate salts or from organic sources such as phospholipids, phosphoproteins, and phytins. Phospholipids and phosphoproteins contain phosphate in a readily available form, but phytate phosphate is not digestible by swine without the addition of phytase enzyme to the ration. Addition of this enzyme greatly increases the availability of phosphorus from phytates. As most of the phosphorus present in grains and oil seed meals is in the phytate form, addition of phytase to typical swine diets will significantly reduce the level of inorganic phosphorus supplementation required.

Environmental temperature will affect swine productivity. Animals need to be able to maintain their body temperature in the face of environmental temperature challenges. Two management approaches are important for helping them accomplish this:

- control the ambient temperature for the animals;
- adjust the energy and nutrient density of the feed as weather conditions vary to allow for animals to adjust nutritionally to those weather conditions (Table 4.5).

For swine raised in ambient temperature housing such as hoop barns, the following feed adjustments in response to temperature changes are expected to be necessary.

When temperatures fall below the lower critical temperature, swine will increase consumption to increase body heat production as follows:

- For each degree Celsius below the lower critical temperature, swine in the 25–60 kg weight range will require an additional 80 kcal ME per day or 25 g more feed; swine in the 60–100 kg weight range required an additional 125 kcal ME or 39 g more feed per day (NRC 2012).
- For each degree Celsius below the comfort zone (i.e., 20–23°C; may be as much as 6°C lower for group house sows such as those in hoop structures), gestating sows require an additional 133–229 kcal ME per day per °C or 40–69 g additional feed per °C. At temperatures above 20°C, no energy adjustment is required (NRC 2012).
- It is uncommon for nursing sows to experience temperatures below their LCT. Estimates, however, are that for each °C below LCT, they require an additional 313 kcal ME (NRC, 1998).

When the temperature rises above the upper critical temperature, swine will consume less feed to reduce body heat production, particularly the heat of digestion (Collin et al. 2001). That intake reduction varies with the stage of production:

- For grow-finish pigs, most estimates indicate a range of 10%−30% decrease in daily feed intake as temperature increases from 19 to 31°C. Other estimates indicate about 80 g/day reduction in feed intake per °C.
- As gestating sows are limit fed, there is no significant impact of high temperature on their feed consumption.
- For lactating sows, the UCT ranges from 18 to 22°C. Decreases in ME intake are temperature dependent as follows: 3300 kcal ME/day/°C between 18 and 25°C; or 7600 kcal ME/day/°C between 25 and 27°C (NRC 2012).

Under high temperature conditions, the percentage of other nutrients (amino acids, minerals, vitamins) in the ration should be increased such that their intake is maintained in face of reduced total intake of feed. It is also advisable to reduce the fiber content of feeds during hot weather as it is the primary source of digestive heat production. In addition, fat may be added to rations during periods of heat stress as it will increase the energy density of the ration without raising the digestive heat increment (NRC 2012).

Another important factor affecting the nutrient composition of rations is the production stage of the animals. Sow feeding after weaning and during gestation is based on the body condition targets, and on the stage of production. The recommended sows' rations are: *1. Flush diet:* fed to the sows from weaning to mating, to stimulate ovum development. *2. Gestation diets:* fed to sows after insemination, or to older parity sows in the first 12 gestation weeks (between day 5 of pregnancy till day 84 of gestation) for backfat stimulating diet, i.e. with lower amino acids to energy ratio. For the late gestation time (last 3 weeks, till day 110) the diet has to be higher in amino acid to energy ratio, in order to enhance piglet birth weights, especially young sows (<3rd parity). 3. *Transition diet:* fed to sows during the transition period between gestation and lactation, from day 110 of gestation till 3 days after farrowing. 4. *Lactation diet:* fed to the sows for the duration of the lactation period, in order to maximize feed intake and milk yield during lactation (www.topigsnorsvin.com). Rations for different stages of production must be balanced for that group. Tables 4.4 and 4.5 give the nutrient requirements for various production stages.

The suggested levels of nutrients may be adjusted depending on individual production performance goals and environmental and management conditions. Digestible energy requirements listed in the table are calculated for the heaviest group within each class of swine. Because of the need for animals to adjust energy intake because of changes in ambient temperature, the following guidelines are useful for hoop barn management systems:

- use self-feeders and allow pigs to adjust their own intake relative to environmental conditions at that time (Fig. 4.2);
- adjust nutrient levels for the growth stage swine are in;

FIGURE 4.2 Feeding area of a finishing barn.

Swine barn of Wisconsin University.

Credit: Photo by Gary Onan, 2007, River Falls, Wisconsin, USA.

- for breeding sows, adjust nutrient density based on temperatures and body condition;
- in cold seasons, feed intake will voluntarily increase and allow for more body heat production;
- in warm seasons, feed intake will drop and so nutrient density within the feeds must be increased.

4.3 Body condition

For sows housed in ambient temperature housing such as hoop structures, allowed feed intake must reflect not only the temperature changes but also the body condition of the animals. Body condition refers to the general degree of fat present in the animal's body at a given time. Body condition reflects the suitability of the animal to meet the physiological demands of the production stage she is in. Body condition is often assessed using a numerical scoring system (*BCS* or *Body Condition Score*). The numerical values are assigned based on an assessment of the amount of fat and muscle tissue located at specific points on the body through subjective (visual) or objective measures (ultrasound of fat thickness). Those locations include the hip, tail head, spine, and ribs as indicated in Table 4.6.

Assessment at each site is performed visually, by palpation, and/or by use of objective measurements of fat thickness (ultrasound or backfat probes). A score between 0 and 5 is assigned for each location, and the final *BCS* is the average of the individual site scores (Sas and Hutu, 2004). Both emaciated or thin and obese sows or gilts will often experience reproduction failure and are not suitable breeding animals.

Body condition scoring is typically performed at various stages of the sow's production cycle such as at weaning and then again at 30, 60, and 90 days postbreeding. Target scores consistent with good production would be 2 at breeding and 3 during gestation with a goal of achieving 3 at farrowing time (a minimum of 80% of sows should reach a score of 3 by farrowing and maximum 20% can be scored at 2.5). Sows may lose 2—4 mm of back fat during a lactation period, but they should not fall below 14 mm, the minimum level of reserves required for a sow to successfully rebreed (Johnson et al., 2006).

During gestation then (days 14—98 postbreeding), feed intake levels should be adjusted such that sows can achieve the target *BCS* at farrowing. Most animals will need extra feed to add body reserves of fat during gestation, but some may require more restricted levels of intake to lose some body fat. At 98 days of gestation, feed allowance for sows can be increased to meet the higher nutrient requirements of the rapidly growing fetuses as well as those for ensuing milk production. At farrowing time, feed allowance may be reduced for a day or two to reduce udder congestion or edema.[3] A typical feeding schedule for days 14—98 of gestation is outlined in Table 4.7.

4.4 Animal welfare

Ewing et al. (1999) describe welfare as the concept of achieving a balance between the animal's instincts and its environment. Most authors indicate that welfare can be described by the degree to which an animal is able to achieve a number of criteria— the so-called "*freedoms.*"

---

[2] The ultrasound operator should palpate the area at the thoracic-abdominal interface until the last rib is found. Moving upward, the transducer should be placed 6 cm off the midline of the sow's spine.

[3] For a typical corn-soy ration with 2985 kcal ME and 14% crude protein, feed should be allowed ad lib for the first 2 weeks following weaning and again after week 14 of gestation. During the 12 intervening weeks, feed allowance per animal will depend on *BCS*. Feeding level for a sow with *BCS* = 1 would be 3—3.5 kg per day; for *BCS* = 2 feed allowance would be 2.5—3 kg per day; for *BCS* = 3, level should be 2—2.5 kg per day. At farrowing time feeding level may be cut to 1.5—2 kg per day if udder edema is a problem.

Table 4.6 Criteria for assessment of sow body condition scores (BCS) and correlation with back fat depth by ultrasound measurements (BF).

Scoring	0	1	2	3	4	5
Picture						
Category **General appearance**	**Emaciated** Visually thin, with hips and backbone very prominent and no fat cover over hips and backbone	**Thin**	**Moderate** The hip bones and backbone are easily felt without any pressure on the hands	**Good** It takes firm pressure with the palm to feel the hip bones and backbone	**Fat** It is impossible to feel the bones at all even with pressure on the palm of the hands	**Obese** It is impossible to feel the hip bones and backbone even by pushing down with a single finger
Spine	Prominent, sharp spinous processes	Prominent spine. Fat cover at 10th rib	Spinous processes not visible but easily palpated	Spine detectable only by firm palpation	Spine not palpable	Midline slightly sunken
Ribs	Very prominent distinct ribs	Ribs less prominent but still visible	Visible or easy palpable covered ribs	Ribs not visible and detectable only with firm palpation	Impalpable ribs	Ribs covered in a thick fat layer
Tail	Deep cavity around tail head	Obvious cavity around tail head	Cavity around tail head filled	No cavity around tail	Tail into fatty tissue	Tail deeply sunk into fatty tissue
Hips	Very narrow. Pelvic bones with sharp edges. Deeply sunken sides	Narrow. Pelvic bones prominent. Sunken sides	Pelvic bones covered and not visible but easily palpated. Slightly sunken sides	Pelvic bones detectable only by firm palpation. Full sides	Pelvic bones not palpable. Full rounded sides at area limits	Thickness of fat layer impossible to assess by even forceful palpation
BF—depth[2] in P2 (mm)		<15	15–18	18–20	20–23	>23

Compilation and adaptation after Coffey, R. D., Parker, G. R., Laurent, K. M., 1999. Assessing Sow Body Condition. Lexington, KY. University of Kentucky College of Agriculture Cooperative Extension Service. ASC-158. Available at: www.ca.uky.edu/agc/pubs/asc/asc158/asc158.pdf, Sas, E., Hutu, I., 2004. Zootehnie - Lucrări Practice, Eurostampa (Ed.), Timisoara, Whitney, H. M., Lactating Swine Nutrient Recommendations and Feeding Management, 2010, PIG 07-01-12, http://porkgateway.org/resource/lactating-swine-nutrient-recommendations-and-feeding-management/.

Table 4.7 Feed requirements for gestating sows and gilts with varying *BCS*.

| BCS | Feed adjustment below or above base ration for days 5–100 of gestation | | | |
	Sow (>160 kg)		**Gilts (>120 kg)**	
1—emaciated	+0.7 to 1.0	kg to base ration	+0.5 to 0.9.0	kg to base ration
1.5	+0.45		+0.35	
2—moderate	+0.35		+0.25	
2.5	+0.25		+0.15	
3—good	Base ration designed to meet the nutrients requirements of an average condition sow based on the environment and production level of the herd			
3.5	−0.25	kg from base ration	−0.15	kg from base ration
4—fat	−0.35		−0.25	
4.5	−0.45		−0.35	
5—obese	−0.55		−0.45	

Adaptation after Foxcroft, G. R., Aherne, F. X., 2000. Management of the gilt and first parity sow: parts I–VI. In: Proceedings of the VII International Symposium on Swine Reproduction and Artificial Insemination, Iguacu, Brazil for feed with metabolizable energy density of 3150 kcal/kg (Luca, 2000).

4.4.1 Liberties for swine

Most authors as outlined in the bibliography would agree that swine production using hoop barn structures for housing ensures that the animals are afforded the *five freedoms* described below:

1. Free access to feed and water: If feed and water are supplied in adequate quantity and in the case of feed with adequate nutrient density and ratios, this freedom will be met. Specifically, the swine manager must assure that water supplies are clean, have adequate flow rate, and are not allowed to freeze. In addition to nutrient density and availability suitable for the production stage of the animals, the manager must also make sure that feed is fresh, palatable, and frequently replenished. As a consequence of exposure to bedding materials, swine in hoop structures will often consume the bedding. This allows for adequate gut fill and diminishes hunger[4] (Pond, 1981), which has important positive effects on swine behavior (Huenke and Honeyman, 2001), particularly that of gestating sows (Honeyman and Roush, 1999).

2. Minimal comfort in the habitat: Hoop barns compare favorably with conventional structures in that they offer similar feeding space and increased resting space. There is, however, more cold temperature stress in hoop barns than in

---

[4] Current breeds and hybrid lines of swine have been bred for rapid growth and high feed consumption and as such are chronically hungry and therefore stressed. Gestating sows are even more stressed as they are typically limit fed (2–3 kg feed per day - 30%–50% of ad libitum intake) to control body condition.

conventional ones, but this problem is addressed through proper nutrition as described in Section 4.2.

Animals housed in hoop barns typically experience less heat stress in summer than in conventional structures. Hoop barns with the tarp covering meet all requirements for sheltering the animals from other adverse weather events.

3. Adequate veterinary health care and freedom from injuries: Swine housed in groups are prone to injury from social aggression (Bhend and Onan, 2003). Proper sizing and composition of groups can greatly reduce this problem. The larger allowable space per animal in hoop structures also helps reduce injury levels.

4. Prevention of pain and undue stress: Hoop barns allow for significant freedom of movement, low levels of harmful manure gases, adequate lighting, and reasonable control of temperature, and therefore relative comfort for the animals housed there-in.[5] Animals housed in hoop structures have the opportunity to alter their own environment, which significantly reduces stress. This is typically not the case in conventional housing. With the presence of bedding materials, animals can build nests to alter the local temperature environment and to offer areas of escape from aggression. The large size of the group coupled with the ability to hide and to express behaviors such as rooting reduces overall aggression significantly.

5. The ability to express natural behavior: Raising animals in deep bedded management systems also offers the opportunity for more expression of natural behaviors such as rooting and exploration (Fig. 4.3). This therefore reduces frustration and boredom and results in decreased stereotypies such as tail biting. According to Lay et al. (2000), animals reared in hoop barns enjoy a high welfare status with a 12-fold decrease in abnormal behaviors compared with animals raised in conventional barns. Hoop barn animals are more playful and exhibit less frequent stress and lower stress intensity as indicated by body fluid cortisol levels.

Furthermore, reproductive behavior is enhanced in hoop barns with better manifestation of estrus and higher estrus rates among sows newly introduced to a group.

4.4.2 Group rearing strengths and weaknesses

Group housing on deep litter bedding has its strengths and weaknesses (Honeyman et al. 1997). In addition to ensuring the freedoms described above, improved animal welfare in hoop systems is manifested by improvements in

- reproductive performance,
- longevity of animals within the breeding herd,
- morbidity, mortality, and frequency of necessary euthanasia.

---

[5] Compared to sows housed in gestation crates, sows housed in deep litter bedding have a reduced neutrophil count but an increased lymphocyte count.

FIGURE 4.3 Animals in hoop structure—exploratory behavior stimulated by straw and human-animal interaction.

First group of pigs in Swine Experimental Unit.

Credit: Photo by Ioan Hutu, Swine Experimental Unit, 2014, Timisoara, Romania.

Honeyman et al. (1997) also suggest that deep bedded management systems exhibit some weaknesses that conventional gestation rearing systems do not such as:

- More aggression during the time social hierarchy is established within the group. This is especially exacerbated with the introduction of new sows into a group. Maintaining stable groups reduces such aggression.
- More aggression during feeding: This problem can be addressed by individual feeding stalls or by the use of electronic feeders (which, however, require significant capital investment).

References

Bhend, K.G., Onan, G.W., 2003. Injury Levels of Sows in Gestation: Stalls vs. Group Housing. American Society of Animal Science Midwestern Branch (Abstr.) p 88.

Brent, G., 1986. Housing the Pig. Farming Press, Ipswich, UK.

Burlacu, G., Cavache, A., Burlacu, R., 2002. Potentialul productiv al nutreturilor si utilizarea lor. Editura Ceres, București.

Brumm, M.C., Dahlquist, J.M., Heemstra, J.M., 2000. Impact of feeders and drinker devices on pig performance, water use, and manure volume. Swine Health Prod. 8, 51–57.

Burlacu, R., Grossu, D.V., Marinescu, A.G., Burlacu, G.M., 2005. Modelarea matematica a proceselor de metabolism energetic si proteic la suine. INBA, Balotesti. Cartea Universitara, Bucuresti.

Carr, J., 1988. Garth Pig Stockmanship Standards.

Close, W.H., 1981. The climatic requirements of the pig. In: Clark, J. (Ed.), Environmental Aspects of Housing for Animal Production. Butterworths, London, pp. 149—166.

Close, W.H., 1987. The influence of the thermal environment on the productivity of pigs. BSAP 11, 9—24.

Coffey, R.D., Parker, G.R., Laurent, K.M., 1999. Assessing Sow Body Condition. University of Kentucky College of Agriculture Cooperative Extension Service, Lexington, KY. ASC-158. Available at: www.ca.uky.edu/agc/pubs/asc/asc158/asc158.pdf.

Crăiniceanu, E., Matiuti, M., 2006. In: Nutritia Animalelor. Brumar, Timişoara.

Collin, A., Milgen, J., Dubois, S., Noblet, J., 2001. Effect of high temperature and feeding level on energy utilization in piglets. J. Anim. Sci. 79, 1849—1857.

Dinu, I., 1978. In: Ceres (Ed.), Influenta mediului asupra productiei la porcine. Bucuresti.

Drăghici, C., 1991. In: Ceres (Ed.), Microclimatul adăposturilor de animale şi mijloacele de dirijare.

Ewing, S.A., Lay, D.C., Borell Jr., E., 1999. Farm Animal Well-Being. Prentice Hall, Upper Saddle River, NJ.

Foxcroft, G.R., Aherne, F.X., 2000. Management of the gilt and first parity sow: parts I—VI. In: Proceedings of the VII International Symposium on Swine Reproduction and Artificial Insemination, Iguacu, Brazil.

Hafez, E.S.E., 1993. Reproduction in Farm Animals, sixth ed. Lea & Febiger, Philadelphia.

Holden, P., Ewan, R., Jurgens, M., Stahly, T., Zimmerman, D., 1996. ISU Life Cycle Swine Nutrition PM-489. Iowa State Univ, Ames, IA, p. 17.

Holmes, C.W., Close, W.H., 1977. The influence of climatic variables on energy metabolism and associated aspects of productivity in pigs. In: Haresign, W., Swan, H., Lewis, D. (Eds.), Nutrition and the Climatic Environment. Butterworths, London, pp. 51—74.

Honeyman, M.S., Roush, W.B., 1999. Supplementation of midgestation swine grazing alfalfa. Am. J. Alter. Agric. 14 (3), 103—108.

Honeyman, M.S., Harmon, J., Lay, D., Richard, T., 1997. Gestating sows in deep-bedded hoop structures ASL-R1496 Management/Economics Iowa State. In: 1997 ISU Swine Research Report, AS-638. Dept, of Animal Science, Iowa State University, Ames.

Huenke, L., Honeyman, M.S., 2001. Fecal Fiber Content of Finishing Pigs in Hoop Structures and Confinement. ASL-R1779. Swine Research Report AS-646. ISU Ext. Serv., Ames, IA.

Johnson, C., Stalder, K., Karriker, L., 2006. Sow Condition Scoring Guidelines. National Hog Farmer. Prism Business Media, Overland Park, KS, 66212-2216. No. 42 in a series. 51:7-9.

Kopinski, J., Willis, S., Spencer, A., 2005. Nutrient Needs and Diet Guidelines for Pigs. DPI&F. http://www.dpi.qld.gov.au.

Kruger, I., Taylor, G., Roese, G., Payne, H., 2006. Deep-litter Housing for Pigs, PRIMEFACT 68. NSW Department of Primary Industries, State of New South Wales. http://www.dpi.nsw.gov.au.

Lay Jr., D.C., Haussman, M.F., Daniels, M.J., 2000. Hoop housing for feeder pigs offers a welfare-friendly environment compared to a non-bedded confinement system. J. Appl. Anim. Welf. Sci. 3 (1), 33—48.

Luca, I., 2000. In: Marineasa (Ed.), Curs de alimentatia animalelor. Timişoara.

National Research Council (NRC), 2012. Nutrient Requirements of Swine. National Academies Press, Washington, DC, USA.

National Research Council (US), 1981. Subcommittee on Environmental Stress. Effect of Environment on Nutrient Requirements of Domestic Animals. National Academies Press

(US), Washington (DC). Swine. Available from: https://www.ncbi.nlm.nih.gov/books/NBK232319/.

Pond, W.G., 1981. Limitations and opportunities in the use of fibrous and by-product feeds for swine. In: Proc, Distillers Feed Conference 36 (April 2): Cincinnati, OH.

Sällvik, K., Walberg, K., 1984. The effects of air velocity and temperature on the behaviour and growth of pigs. J. Agric. Eng. Res. 30, 305–312.

Sas, E., Hutu, I., 2004. In: Eurostampa (Ed.), Zootehnie - Lucrări Practice. Timişoara.

Taylor, G., Roese, G., Brewster, C., 2006. Water Medication for Pigs, PRIMEFACT 108. NSW Department of Primary Industries, State of New South Wales. http://www.dpi.nsw.gov.au.

Verstegen, M.W.A., van der Hel, W., 1976. Energy balances in groups of pigs in relation to air velocity and ambient temperature. Pub. Eur. Ass. Anim. Prod. 15, 347–350.

Verstegen, M.W.A., Close, W.H., 1994. In: Cole, D.J.A., Wiseman, J., Varley, M.A. (Eds.), Principles of Pig Science. Nottingham Univ. Press, Nottingham, UK, pp. 333–353.

Yousef, M.K., 1985. Thermoneutral zone. In: Yousef, M.K. (Ed.), Stress Physiology in Livestock, vol. I. CRC Press, Boca Raton, FL, pp. 47–54. https://topigsnorsvin.com. White Paper Sow Feeding - Sow feeding after weaning and during gestation. Available at: http://www.finnpig.fi/wp-content/uploads/2017/05/Sow-feeding_2-feeding-weaning-gestation.pdf.

Swine production management in hoop barns

5.1 Overview of production cycles in hoop barns

The techniques for management of swine in alternative systems have been described for two Swedish variations, the Thorstensson and Vastgotmodellen systems (Algers 1991), and for a variety of approaches to raising hogs in hoop structures in the United States and Canada (Halverson, 1991a, 1991b, 1994; Honeyman 1995). The common characteristics of these systems are the raising of swine in large groups and the use of deep bedded packs within those group pens for manure management.

A complete cycle of farrowing, weaning, and finishing in hoop barns requires two or three different types of structures be used for the six basic production stages as presented in Fig. 5.1. Several cycles per year may be run in each structure depending on the stage of production it is designed for.

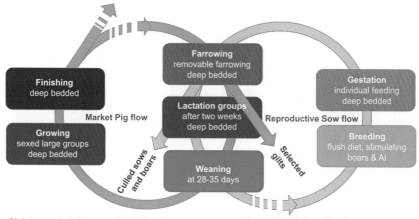

FIGURE 5.1 Production and breeding cycles in deep bedded hoop barns.

Adapted from flow diagram of Thorstensson version of Swedish pig production.

Alternative Swine Management Systems. https://doi.org/10.1016/B978-0-12-818967-2.00005-8
Copyright © 2019 Elsevier Inc. All rights reserved.

For breeding sows, one reproductive cycle will require approximately half a year as broken down here:

• Breeding	6 days (typical)
• Gestation	110 days
• Farrowing/lactation	28–38 days
Total for one cycle	**144–154 days**

To bring a gilt with current hybrid genetics into production requires about 316–337 days apportioned as follows:

production stage	duration	age/weight
• Suckling/nursery	50 days	50 days/18 kg
• Growing phase	110 days	160 days/100 kg
• Acclimation and estrus stimulation	60 days	220 days/130 kg
• Puberty		
• Breeding, second or third estrus	21–42 days	241–262 days/ 137 to 155 kg
• Gestation	115 days	Gain of 65 kg
• Farrowing (after second or third estrus)		202–220 kg
Total time to first farrowing for a gilt =		**356 to 377 days**

Total time for finishing market hogs from farrowing until sale is about 180 days broken down as follows:

• Suckling/nursery	50 days
• Grow/finish	120–140 days
Total time for one finishing cycle	**170–190 days**

Typically, 2.2 to 2.4 farrowing cycles can be completed for each sow group per year. The makeup of the sow groups will affect overall fertility of the group. It is common to cull sows at weaning time that are reproductively unfit or are performing below production goals.

The optimal composition of the sow herd at any given time is 17% gilts, 15% primiparous sows, and then about 1% fewer for each parity up through the sixth. Therefore, the sixth parity animals should comprise about 10% of the herd. Older animals constitute a relatively small percentage as many managers cull all but the very best animals after six parities. Seventh parity sows may comprise only about 5% of the herd and eighth parity only about 3% of the herd. Deep litter bedding production systems often result in sow longevity that allows for six or seven parities (Halverson et al. 1997).

For grow-finish groups, a reasonable goal is 2.5—2.8 cycles per housing unit per year depending on input and output weights. If the structure is used as a wean-finish unit, then 2.0—2.3 cycles per year would be desirable.

5.2 **Breeding management**

Management of gilts and of sows postweaning must be designed to promote expression of estrus and limit aggression during the breeding period as well as the following 30 days so that stress is reduced, and there will be fewer early embryonic losses.

5.2.1 **Breeding hoop structure design**

During breeding, it is important that the manager be able to stimulate puberty in gilts, flush gilts, control boar movement during heat detection, allow for animals to freely express estrus behavior, and control movement of sows during insemination. The hoop barn must be designed to facilitate these requirements. Some of these design features are described here:

1. Resting area (bedded) per head: $6.0\ m^2$ for boars and $> 2.2\ m^2$ for sows
2. Animal movement alleys: 0.90—1.25 m wide surfaced with rough concrete or a mat
3. Feeding area floor: Nonslippery concrete

The inner region of the hoop barn must be divided into service alley, sow pens, and boar pens.

Boar pens may be located in the same structure as the sows or in an adjacent structure. Figs. 5.2 and 5.3 offer two variants for laying out a breeding area in a hoop barn. One approach has pens for sows on one side of the service alley and boar pens as well as a general area for miscellaneous activities such as weighing, veterinary treatments, or isolation located on the other side. Hoop dimensions as indicated in Fig. 5.2 are 22.0 m length, 11.2 m width, and 5.6 m height. This provides $6.14\ m^3$ air/sow and $20\ m^3$ air/boar for a herd consisting of 40 sows and 4 boars.

The service alley should run the entire length of the structure and be 0.90—1.25 m wide, which will allow one person plus the heat detection boar to walk alongside each other.

The alley surface should be rough concrete at a level 0.45 m higher than the animal resting area. It should have a 1.5% slope from one end of the barn to the other with no cross-sectional slope required. The miscellaneous work area may be used for weighing or feed storage where hand feeding is performed.

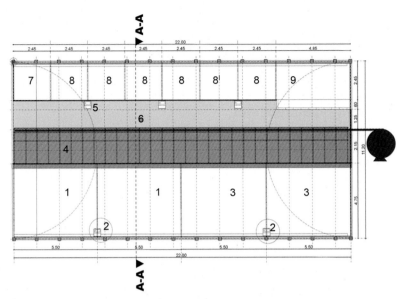

FIGURE 5.2A Breeding structure for 40 sows in group pens. Floor plan.

1. Sow group pens; 2. water fountains; 3. gilt group pens; 4. feeding stalls; 5. water fountains and feeders for boar pens; 6. service and animal movement alley; 7. weighing pen; 8. pens for young and adult boars (8′); 9. multipurpose space—pen for subgrouping of gilts or for straw storage; 10. automatic feeding system with outside feed bin and auger.

Redraw after Harmon, J. D., Honeyman, M. S., Kliebenstein, J. B., Richard, T., Zulovich, J. M., 2004. Hoop Barns for Gestating Swine. Rev. Ed. AED44. MidWest Plan Service. Iowa State Univ., Ames, IA. p. 20.

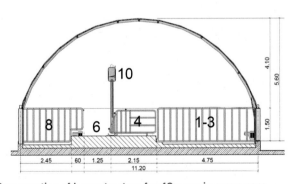

FIGURE 5.2B Cross section of hoop structure for 40 sows in group pens.

1. Sow group pens; 2. water fountain; 3. gilt group pens; 4. feeding stalls; 5. young boar water fountains and feeders; 6. service animal movement alley; 8. young and adult boars (8′); 10. automatic feeding system.

Adapted and redrawn after Harmon, J. D., Honeyman, M. S., Kliebenstein, J. B., Richard, T., Zulovich, J. M., 2004. Hoop Barns for Gestating Swine. Rev. Ed. AED44. MidWest Plan Service. Iowa State Univ., Ames, IA. p. 20.

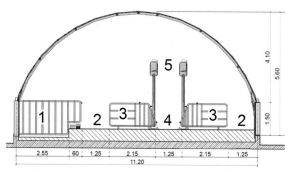

FIGURE 5.3 Cross section of hoop structure for 64 sows in gestation crates.

1. Pens for young and adult boars; 2. service alleys; 3. two rows × 32 gestation crates (2.15 × 0.60 m), with flush gutters; 4. service and animal movement alley for boars; 5. automatic feeding system with outside bin and drop box for each sow.

Adapted after Harmon, J. D., Honeyman, M. S., Kliebenstein, J. B., Richard, T., Zulovich, J. M., 2004. Hoop Barns for Gestating Swine. Rev. Ed. AED44. MidWest Plan Service. Iowa State Univ., Ames, IA. p. 20.

Sow pens are divided into two general areas—one for resting which is deep bedded and one for feeding. The feeding area should have feeding stalls (as described in Section 3.3) to prevent undue aggression during eating. Each sow pen should be approximately 6.90m × 5.50 m, which will allow 3.79 m²/sow (Fig. 5.2). Other factors to consider are as follows:

- Pens should have portable dividers (gates) so that size of pens can be adjusted depending on the population of the sow groups (typically 10 sows/group).
- The front gate of the feeding stalls should be designed to swing open to allow sows access to, or exit from, a given pen.
- The resting area should be 4.75 m wide and the floor should slope 0.5%–1.5% from one end of the hoop to the other.

The feeding area floor should be concrete, and the resting area floor can be packed clay or concrete. As indicated in Fig. 5.2, water access may be provided through addition of water to feed pans during feeding, but the primary source of water would be through freeze-proof fountains located on concrete pedestals in the resting area of the hoop along its wall.

Boar pens should be designed to accommodate one animal each—about 3.0 m × 2.5 m. Each pen has a gate that allows the boar to exit into the service alley for estrus detection or to enter breeding pens for natural service. Boar pens should have sloped concrete floors similar to the service alley. Pens for natural service should be a minimum of 10 m² and have a bedding covered concrete floor.

The bedding storage space (item 9 in Fig. 5.2) can be a temporary pen for mixing of new groups of sows before movement into the normal sow pens.

Fig. 5.3 shows a design for a hoop barn where sows will be kept in gestation crates during the breeding cycle rather than in deep bedded pens. The hoop structure's outside dimensions are the same as for the design in Fig. 5.2, namely 22.0 m long by 11.2 m wide and 5.6 m high. All the other building components are basically the same except that there will be more service alleys and gestation stalls are used in lieu of feeding stalls. There will be three service alleys in this design with the same dimensions and slope as for the previous design. The three longitudinal alleys should be connected by cross alleys, which are slightly wider—1.4 m versus the 1.0 m width of the main alleys. The gestation stalls are typically 2.2 m × 0.6 m in dimension. Because the entire sow area of the barn is filled with stalls, the building will accommodate a higher population of animals—66 sows. With this management system, sows are maintained in this structure until day 28 of gestation and then transferred to a deep bedded group pen structure such as that described in Fig. 5.2.

This type of barn will contain two rows of stalls with concrete floors (no bedded areas at all), and manure removal can be manual or steel grate-covered gutters can be built into the floor at the rear of each row of stalls and the manure removed from them using a water flush system. Each animal will have individual access to feed and water—both of which may be provided in the feed trough that typically runs across the front of each row of stalls. In addition, the operator may choose to install water nipples in each stall. Feed delivery may be manual but is most commonly done by an automatic feed distribution system identical to those used in conventional gestation units. Boars in this unit may be maintained in deep bedded pens identical to those described for the design in Fig. 5.2.

A third breeding barn design is illustrated in Fig. 5.4. The sow housing structure in this case will be 36.6 × 14.0 × 7.0 m dimension. This portion of the unit will be divided into a resting area with deep litter bedding, a feeding area, and service alleys allowing access to the breeding pens.

The service alley, 0.9–1.2 m wide, will be along one of the walls of the structure and will provide space for stockpeople to walk by or for movement of sows or boars for breeding. Boar pens will be located outside the perimeter of the sow barn in this design.

There will be six sow pens, each providing space for 10 sows. Each pen will have a resting area of 10.0 × 6.0 m with a packed clay or concrete floor. This area will be bedded. The feeding area for each pen is located on a concrete platform 0.40 m higher than the floor of the resting area. The feeding platform is 4.25 m wide. Feed may be delivered using conventional automatic drop feeders. One water fountain will be located at the dividing gate for each pair of pens (Fig. 5.4). The feeding area in each pen may have a pair of 3.0 m gates installed to confine sows to the feeding area for the purpose of veterinary treatments, vaccinations, or pregnancy detection.

Boar pens in this design are located in a separate structure outside the perimeter of the sow structure but located adjacent to it. Each of the four boar pens is 4.5 × 3.5 m providing a total of 15.75 m^2 per boar. The pens face an alley 2.5 m wide.

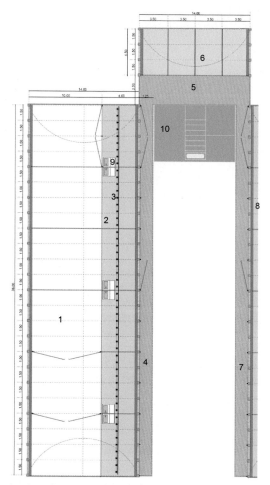

FIGURE 5.4 Floor plan for a heat detection and breeding center located between two gestation units.

1. pens for 10 sows/gilts; 2. feeding stalls; 3. individual drop feeders and delivery system; 4 and 7. access and animal movement alleys for sows; 5. cross access alley to boars; 6. pens for boars; 8. a second barn for sows; 9. water fountains; 10. area for estrus detection or boar exposure for gilts and sows (see Fig. 5.5).

Adapted and redrawn after Harmon, J. D., Honeyman, M. S., Kliebenstein, J. B., Richard, T., Zulovich, J. M., 2004. Hoop Barns for Gestating Swine. Rev. Ed. AED44. MidWest Plan Service. Iowa State Univ., Ames, IA. p. 20.

Opposite this alley an optional estrus stimulation/detection area may be constructed (item 10 in Fig. 5.4 with details in Fig. 5.5). This optional breeding area is useful for stimulating puberty in gilts over 150 days of age by providing boar contact and is useful for estrus detection of sows by offering the opportunity to easily expose them to multiple boars. The sow pens are located on either side of this facility

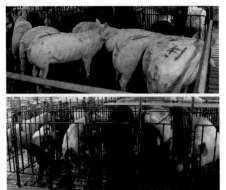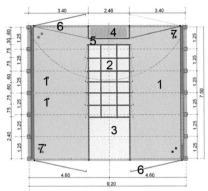

FIGURE 5.5 Detailed floor plan for boar exposure area from Fig. 5.4.

Right: 1. estrus detection pens for 15–20 gilts; 1′. breeding pens created by splitting estrus detection pens; 2. 4 to 6 pens for boars (60–75 cm × 1.80–2.40 m, 1.5 m high); 3. pen for females in heat (1.80–2.40 × 1.2–1.5 m, 1.0 m high), 3 out of 4 gates movable; 4. scale pen (1.80–2.40 × 0.61 × 1.0 m); 5. passage for stock person (1.8–2.4 × 0.25 to 0.35 × 1.0 m); 6. access gates for sows; 7. posts for a refuge area for nonestrus sows; 7′. space that results from removal of posts from pen for mating. Left: A set of boars in the stalls of the boar exposure area facing each of the pens (left down) and gilts on one side of the stimulation and detection area of the boar exposure area (upper left).

Redrawn and modified after Beltranena, E., Patterson, J., Willis, J., Foxcroft, G., 2005. The boar exposure area of the GDU (Gilt Development Unit), Western Hog J. 27 (1): 29–31 and Levis, D. G., Gilt Management in the BEAR System, Pork Information Gateway, Factsheet, University of Nebraska, 2015. http://porkgateway.org/wp-content/uploads/2015/07/gilt-management-in-the-bear-system1.pdf (right). Photograph courtesy of Jennifer Patterson, Swine Research & Technology Center, University of Alberta, Edmonton, Alberta, Canada.

as indicated in Fig. 5.5. These are fixed with swinging gates at each end allowing easy access to service alleys from the main sow housing hoop structure. These sow pens may be provided with internal divider gates hinged to the center post to allow for subdivision into four pens if needed. In each corner of the sow pens, two posts are installed, which create a refuge for nonestrus sows needing escape from an aggressive boar. The boar pens in the center of this area are designed to be of variable size (60–85 cm in width) with moveable gates 1.2–1.3 m in height. Boars can be positioned to face either of the two sow pens, and gates are provided at each end of the boar pen to allow access into either sow pen. Area 3 in Fig. 5.5 may also be used as a breeding pen as needed.

An optional scale pen (with sides 1.0–1.1 m high) may be located opposite the boar pens. This is a useful adjunct for weighing gilts to determine whether they have reached adequate size for breeding.

Skilled staff may, however, be able to make this judgment effectively without the necessity of a scale. Between the scale pen and the boar pens, a narrow alley 30–40 cm wide for stockpersons to view sow and boar mating activity is provided.

Once estrus detection is completed, the sow pens may be subdivided as indicated in the previous paragraph into four smaller pens for actual natural service or artificial insemination. Each pen will then be $10.8\,m^2$ ($3.6 \times 3.0\,m$—see right half of Fig. 5.5). The smaller pen allows for better control of sow movements during mating. In addition, the refuge posts in the corners may be lifted out of their floor sockets and removed during actual insemination procedures. Sows in heat may be moved to area 3 in Fig. 5.5 for insemination as well.

5.2.2 Breeding management

Management of breeding in a sow herd includes such activities as gilt selection, movement and grouping of breeding females, estrus detection, and culling of unproductive animals.

Acquisition of new breeding animals (gilts) into the herd is a significant potential source for introduction of new diseases. Because of this, many herds produce their own replacement gilts internally, oftentimes using artificial insemination to bring in new genetics. If gilts must be purchased from outside sources, it is advisable that a continuing relationship be established with a single herd, which maintains a high health status.[1]

Mixing of gilts from several sources is strongly discouraged. Once new gilts are introduced to the farmstead, they should be vaccinated, quarantined, and acclimated,[2] for 30—60 days under the supervision of a veterinarian (Baker 2004).

Selection of replacement gilts from within the herd is an ongoing process with keep/cull decisions made at various stages of the gilt's life (Table 5.1). The process begins with the birth of the gilts and proceeds through weaning, growing, and puberty with final selections being made at breeding time.

As far as possible, sow groups should be handled on an all-in-all-out basis and remain static.[3] As animals are culled from sow groups, gilts will necessarily need to be introduced to maintain an economically viable group size. New subgroups of gilts may be placed into breeding pens 3—4 days before the introduction of the sows, thus allowing the gilts to settle in and familiarize themselves with the area (Putten and Bure 1997) and therefore be better able to mount a self-defense strategy when the older animals are introduced. Except for new introductions such as this, dynamic grouping is ill-advised as it results in lower reproductive performance

---

[1] Such herds should be free of important swine diseases such as PRRS (porcine reproductive and respiratory syndrome), mycoplasmal pneumonia, Salmonellosis, and others.

[2] To acclimate gilts to the diseases specific to the farm and allow them to build some immunity before breeding, a cull sow may be intermingled with them during the quarantine period.

[3] A static group is one where all sows in the group are managed together, housed together, and all are bred, farrowed, and weaned simultaneously. A dynamic group is one where animals are continuously being added to and removed from the group. For example, all gestating sows might be housed in one large area, and newly weaned sows are being routinely added and sows due to farrow are removed. This is not a very effective way to manage a large herd of sows.

Table 5.1 Decision points for gilt selection.

Time point	Actions at each time point
Birth	• Identifying potential replacement gilts in each litter, noting those with defects (such as umbilical hernias and the like) • Recording birth date and number of piglets in the litter • Identification by ear notching • Equalizing litter sizes by fostering male piglets to sows with small litters (after colostrum intake)
Weaning	• Feed for adequate development and growth • Identify gilts that were initially marked with ear tags or notches: Check nipple number, spacing and shape. Select two to three times as many gilts as the expected need for replacement will be

Near puberty (normally at ages >160 days and weights over 90–100 kg)	• Check five criteria:
	  • Growth that meets production goals: gilts picked for replacements should be in the top 25% of their peer groups for growth and meet minimum production goals; weight by mating time should be 130–140 kg
	  • Optimal body condition: Replacement gilts will be scored (see Table 4.8) for Body condition or have 10th rib fat measurements made; those retained should have met minimum condition goals
	  • Underlines: Gilts must have at least six pairs of functional nipples that are evenly spaced; gilts with more than two inverted or pin nipples should be culled
	  • Development of external genitalia: Gilts with infantile, tipped, or injured vulvas should be culled
	  • Structural soundness: Gilts will be evaluated for skeletal soundness and any with defects such as high backs, straight legs, or fine bone structure should be culled; retained gilts should be from large litters; typically, 30%–40% more animals should be retained at this stage than what expected needs are (Stalder et al. 2005)

Breeding	• Breed those gilts that show earliest estrus as they will be more fertile throughout their lifetime; once needed number has been mated, cull the remainder • Check that identification by ear notches or tags are intact

Adaptation after Sas, E., 1996. Curs de zootehnie si etologie, Ed. Brumar, Timişoara, Hutu, I., 2005. Reproducerea intensivă la scroafe, Ed Mirton, Timişoara, and Stalder, K. J., Johnson, C., Miller, D. P., Baas, T. J., Berry, N., Christian, A. E., Serenius, T. V., 2005. Pocket Guide for the Evaluation of Structural, Feet, Leg, and Reproductive Soundness in Replacement Gilts. Des Moines, IOWA: National Pork Board. Pub. no. 04764.

(O'Connell et al. 2002)—such decreases, however, may be mitigated to some degree by the presence of a boar (Docking et al. 2001).

Stimulation of puberty and estrus detection should begin in gilts at 150—170 days of age (Hutu and Chesa, 2002). Onset of puberty can be advanced by exposure to boars[4] in an area such as that illustrated in Fig. 5.5. Gilts are brought into the sow pens daily and will be in nose-to-nose contact with 4 to 6 boars in the small pens in the middle of the area. Herds people should watch for signs of estrus at this point (flushed and swollen vulvas, ear pinning, mounting, etc.—Fig. 5.6) and if indicated, check for the immobilization reflex (standing heat) by applying pressure to the backs of the animals. Any animals in standing heat may be moved to a breeding pen.

An experienced boar (>10 month of age) may then be introduced into the heat detection pen to determine if any gilts will exhibit immobilization with mounting by the boar. These gilts can also be removed to the breeding pen. After 15—20 min of in-pen boar exposure, gilts in the heat detection pens can be removed to their normal housing area.

All gilts that were removed to the breeding pen may be inseminated if their weight is over 135 kg, back fat thickness has reached 12—20 mm, and they are exhibiting their second or third estrus. The heat detection pens may be used for mating or artificial insemination also by swinging dividing gates and decreasing the pen size.

FIGURE 5.6 Stages of swine sexual behavior.

1. Salivation; 2. tactile and visual contact; 3. sniffing of genital area; 4. sow chasing;
5. chin leaning against lower neck, ear biting; 6. mounting.

Credit: Hutu, I., 2005. Reproducerea intensivă la scroafe, Ed Mirton, Timişoara.

---

[4] Boar contact stimulates and intensifies luteinizing hormone release, follicle growth, and ovulation.

Table 5.2 Ovulation timing as a function of weaning to estrus interval.

Probability of ovulation timing in relation to onset of standing estrus	Weaning to estrus interval			
	3 days (19%)	4 days (54%)	5 days (22%)	6 days (6%)
0–32 h (35%)	22%	28%	52%	73%
32–48 h (55%)	57%	63%	43%	18%
48–64 h (10%)	22%	8%	5%	9%

Adapted from a study by Kemp, B., Soede, N. M., 1996. Relationship of weaning-to-estrus interval to timing of ovulation and fertilization in sows. J. Anim. Sci. 74, 944–949.

Any gilts not exhibiting estrus after 21–23 days of twice daily heat detection may be regrouped and treated with the hormonal adjunct PG600.[5] If animals still fail to exhibit estrus after the hormonal treatment, they should be culled and can still be marketed as finished hogs after a brief period of heavier feeding. Research indicates that with daily heat detection in the presence of multiple boars, 97% of gilts will cycle with an average estrus duration of 55 h (Beltranena et al. 2005).

Estrus detection for sows should begin 3 days postweaning. The procedures used for heat detection are varied.

One can use the same facility and method as used for the gilts. On the other hand, boars may be introduced directly into the sow pens or sows may be allowed into an area near the boar pens. Sows may be moved past boar pens after feeding or boars may be paraded down the service alley in front of feeding stalls after sows complete their feeding (Gonyou 2003). If sows are mated twice, litter size typically increases by 0.3–0.5 piglets (Koketsu and Dial 1997).

Insemination timing is a function of the basic reproductive physiology of the sow. Ovulation occurs at 70%–75% of the total duration of estrus (Table 5.2—most frequent timing is 32–48 h after onset of standing heat—Kemp and Soede, 1996). Ovulation duration itself requires 1–6 h, and ova are fertile for 6–8 h postovulation. Sperm require about 6 h for capacitation and will remain fertile for about 12 h post-insemination. All these facts then determine when insemination should occur. Several different insemination timings are used depending on the length of the weaning-to-estrus interval for any given sow (Fig 5.7 and Table 5.2). In any case, for each approach, fertile sperm will be available for 30–36 h during the period of time when the sow is most likely to ovulate.

---

[5] Use of this brand implies no endorsement of it over others.

[6] Given that 85% of sows ovulate within 44–60 h from onset of standing heat, a third mating may occur as late as 48 h after onset; in such cases the probability is high that quite a number of sows will no longer accept mating, however (Knox, 2002). Do not inseminate sows in late estrus that are no longer exhibiting a good immobilization response. Late inseminations may result in uterine infections.

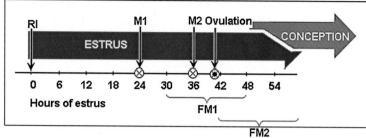

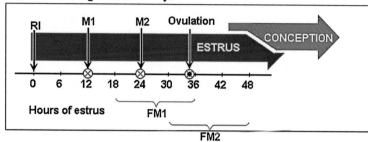

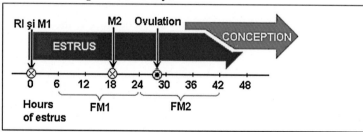

FIGURE 5.7 Mating or insemination timing.

IR, immobility reflex; M1, first mating or insemination; M2, second mating or insemination; FM1, fertile period due to M1; FM2, fertile period due to M2. **Approach A:** for sows showing estrus ≤4 days postweaning, estrus duration is longer and ovulation later—first mating or insemination should be at 24 h after onset of standing heat and repeated in 12–18 h.[6] **Approach B:** for sows entering estrus 5 days postweaning—first mating or insemination 12 h after onset of standing heat and repeated in 12 h. **Approach C:** for gilts and for sows entering estrus ≥5 days postweaning, estrus is short and ovulation early—in which case the first mating or insemination should be done immediately and repeated after 12 h.

Credit: Hutu, I., 2005. Reproducerea intensivă la scroafe, Ed Mirton, Timişoara.

If artificial insemination is used, a boar must still be present during the insemination as he will stimulate release of oxytocin, which will facilitate movement of semen through the sow's reproductive tract to the site of fertilization.

The success of the breeding protocol (conception rate) is dependent on a variety of factors. Effective heat detection and proper insemination timing are two critical factors. Double insemination will raise conception rate as fertile sperm will be available for a longer period of time in the female's tract, thereby increasing the likelihood of their presence during ovulation when ova are most fertile. If semen is introduced too early, sperm fertility will be impaired by the time the sow ovulates. Conversely, if semen is introduced too late, ova fertility will be impaired by the time sperm arrive.

It is advised that for natural service of sows in group housing, hand mating be used. This allows for better control of boar usage rate and exact knowledge of which sires have been mated to which sows and exact breeding dates so that precise expected farrowing dates can be calculated. Matings may be either monospermatic (one boar) or heterospermatic (two or more boars).

Sows that do not exhibit estrus during the first week postweaning may be regrouped for breeding on subsequent cycles. The timing of such regrouping depends on the age and size of the sows (Sas, 1996 as outlined in Table 5.3). Old sows will undergo the first regrouping 12 days postweaning with a second regrouping at 24 days postweaning. If such sows still fail to show heat, they should be culled. Only gilts or young sows would be retained for a third regrouping and then only in the summer months (particularly if less than 75% of gilts exhibit estrus).

Another strategy for synchronizing sows or cyclic gilts is the use of Matrix (Merck Animal Health), a solution of altrenogest (synthetic progesterone) designed specifically for oral administration to female swine. Administration of 6.8 mL/head/

Table 5.3 Regrouping strategy (R-I to R-III) for late estrus females.

Class	R-I	R-II	R-III
Gilts	45 days after initiation of heat detection	25 days after R - I	Only over summer
Small/medium sows	14 days postweaning	24 days after R - I	Only over summer
Large sows	12 days postweaning	24 days postweaning	15 days after entering R-II, cull
Miscellaneous sows[7]	Begin heat detection 21 days after litter removal and then regroup based on above classifications		
Cull sows	Are not checked for estrus		

After Sas, E., 1996. Curs de zootehnie si etologie, Ed. Brumar, Timişoara and Hutu, I., 2005. Reproducerea intensivă la scroafe, Ed Mirton, Timişoara.

---

[7] Miscellaneous sows—that is, sows that had unsuccessful farrowings or had small litters that were fostered off between 2 and 15 days; start heat detection 21 days after litter removal.

day (as a single dose each day) for 14 days will result in estrus occurring 4–9 days following the last dose. Matrix may be topdressed on the animal's daily feed allowance or drenched with a dosing gun. Swine like the taste of Matrix and will readily accept daily drenching.

Sows that fail to return to estrus following weaning will often respond to PG600 (Merck Animal Health) and show heat within 4–5 days following injection of a 5 mL dose.

Culling decisions for sows begin at weaning with sows having reproductive problems or who are poor performers being removed at that time (Hutu, 2004a and 2004b). After breeding time, sows failing to return to estrus may be regrouped or immediately culled. Also, particularly in cold housing management systems, poor body condition may be a criterion for culling (Lammers et al. 2006). Boars can typically be used for 2–5 years before culling. In less intensive mating systems, a cull rate of 30%–40% may be expected, whereas in intensive systems, it may reach 50%–60% per year for boars.

5.2.3 Barn environment

The most frequently used design for breeding barns is illustrated in Fig. 5.2. Feed delivery, ventilation, and other environmental control will be as outlined in Chapter 3. For breeding swine in deep bedded housing, the ideal temperature range is 15–20°C. Optimum air movement is 20–120 m^3 per hour.

Research from Iowa, the United States, indicates that the mean summer temperature in such structures is 26°C, and the maximum temperature achieved is 36°C (Honeyman 1997). About 14–16 hours of light per day at 100–150 lux is desirable for sows during the breeding period. Bedding requirement is about 1.8 kg straw per sow per day (Kruger et al. 2006), and manure clean-out is typically four to five times per year (Honeyman and Kent 1997).

During the summer infertility period, a small amount of oxytocin (4–5 IU) may be added to the semen just before insemination. This has been shown to increase conception rate.

5.2.4 Animal management

Sows may be housed in either individual stalls (as in Fig. 5.3) or large group pens (as in Fig. 5.2; Harmon et al., 2004). During breeding, boars may be housed in the same structure as the sows, although the better approach is to house them in a separate nearby structure as sows will respond with stronger signs of estrus if boars are out of sight and smell range for the time between periods of actual estrus detection.

Older experienced boars can be used more heavily than younger ones (Table 5.4). A typical ratio is one mature boar per 20 sows. Artificial insemination will significantly decrease the requirement for boars as well as speed up genetic progress and is therefore the preferred method for mating sows. For stimulation of puberty onset in

Table 5.4 Recommendations for boar work load.

Boars group	Heterospermatic mating (matings per day/rest days)		Homospermatic mating (matings per day/rest days)
	Average management	Intensive management	
Adult boars over 15 months	1/1	2/1	2–3/3
Young boars 12–15 months	1/2	1/1	2–3/5
Young boars 9–11 months	1/3	–	2–3/6
Young boars 8–9 months	1/7	–	–

After Dinu, I., Hälmăgean, P., Tărăboantă, G., Farkaş, N., Simionescu, D., Popovici F., 1990. Tehnologia creşterii suinelor, Ed. Didactică şi Pedagogică, Bucureşti, Sas, E., 1996. Curs de zootehnie si etologie, Ed. Brumar, Timişoara, and Hutu, I., 2005. Reproducerea intensivă la scroafe, Ed Mirton, Timişoara.

gilts, a group of 3 boars for every three groups of 15–20 gilts is recommended (Beltranena et al. 2005a).

5.2.5 Vaccination

Vaccination protocols should be established to prevent transmission of disease between animals and to some degree prevent transmission of zoonotic diseases between humans and animals or vice versa. It is common at minimum, to vaccinate sows at weaning for parvovirus, leptospirosis, and atrophic rhinitis (Lammers et al. 2006). Boars should be vaccinated every 3 months. Every 6 months, both sows and boars should be vaccinated for Erysipelas (Moga 2001).

5.2.6 Nutrition during the breeding period

Boar nutrition: Boars require a daily intake of 7800 kcal ME, 170 g standard ileal digestible (SID) protein of which 12 g is SID lysine. (This equates to roughly 209 g of crude protein with 14.25 g total lysine.) Boars also require about 18 g Ca, 7.8 g available P, and 2.4 g linoleic acid. In addition, the ration must contain 9500 IU of vitamin A and 104.5 IU of vitamin E.

Typically, boars will receive these nutrients as a once-daily dry mixed feed ration. Usually, 2.2–3.0 kg/boar per day is adequate depending on body condition and level of breeding activity. During periods of no mating activity, feed intake may be cut to 1.5–2.0 kg per day. During summer, feed amounts can be cut by 25%–30% and replaced with 2–3 kg of alfalfa or clover, and in winter, carrots may be used to make up this portion of the ration (Roese and Taylor, 2006a).

Base ration ingredients are of prime importance for meeting the boar's basic nutrient requirements as well as the requirements for normal spermatogenesis. Corn (maize), barley, oats, peas, and soybean or sunflower meal are excellent nutrient sources. Most farms in Romania feed a ration with 2950–3050 kcal DE/kg, 14% digestible protein, 0.75% Ca, 0.5% P, and 0.5% salt (Luca 2000).

These nutrient levels can be achieved with:

- energy feeds: 30%–40% barley; 20%–40% corn; and 10%–20% oats
- protein feeds: 10% peas; 8%–12% sunflower or soybean meal; and 3%–5% feed dross
- mineral feeds: 0.5% salt; 1.0%–1.5% limestone; 0.5%–1.0% phosphates; and 1.0% vitamin/trace mineral premix

Sow and gilt nutrition: Nutrient requirements of sows during breeding are very similar to that of boars. Gilts, however, may respond with a higher ovulation rate if flushed (allowed increased feed intake around breeding time), which may begin 2 weeks before breeding and continue for 2–3 days postbreeding.

Current recommendations for feeding sows during breeding indicate improved performance if sows are allowed ad lib feed intake during the breeding week (field trial data from PIC, International). During the trial, sows in the ad lib group had an average daily feed intake of 4.25–4.50 kg and control animals were limit fed at 2.6 kg/day. The ad lib group experience a wean-to-first service interval 1 day shorter (4.4 vs. 5.3 days), had 5% more sows serviced by 7 days (97.5% vs. 92.8%), and averaged one more pig/litter (13.9 vs. 12.9 total born) on their subsequent farrowing than the limit-fed group. When feeding sows ad lib during breeding, their feed intake must be immediately reduced by about day 8 postweaning to the amount required for proper body condition development during gestation (see Section 5.3.6 for gestation feed recommendations). Excessive feed intake during early gestation typically increases early embryonic loss.

During winter, an extra 1/3 kg of feed for each 5° below 13°C should be given. Successful breeding management in winter depends on providing extra feed and maintaining a high-quality bedding pack (Honeyman et al. 1997).

5.2.7 Reproduction performance

Research has indicated that hybrid sows, including Yorkshire, Landrace, and Hampshire crosses (Honeyman et al. 1997), work well in hoop structure production systems. In Romania, besides these specialized breeds, breeds adapted to the Romanian climate may also be utilized. Such breeds include Duroc, Large White (Hutu, 2005; Sas et al. 2000a, 2000b) as well as native breeds such as Bazna and Mangalitza (Brussow 2005) or the Black Strei type (Hutu, 2005).

For hybrid sows consisting of ¼ Hampshire × ½ Yorkshire × ¼ Landrace raised on deep litter, Honeyman and Johnson (2002) report that postweaning, 92% of sows returned to estrus with a mean weaning-to-estrus interval of 7.5 days. In another trial (Lammers et al. 2006), it was found that sows housed in groups with deep litter bedding had a longer wean-to-estrus interval (6 days) than sows housed in conventional gestation crates (5 days).

Most authors indicate no impairment of reproductive performance due to cold temperatures in hoop structures during winter. Sows and boars will consume more feed, however. High summer temperatures are more of a problem. Beginning at about 2 weeks of exposure to high ambient temperatures, boars, for example, will have reduced fertility, which will not recover to normal for 6–8 weeks following the heat episode. Sows will have fertility indices 10%–20% lower during episodes of high ambient temperatures.

The most critical time when temperature stress impairs sow reproduction is exposure during the first 2–3 weeks of gestation or the last 2–3 weeks. High heat during these times will significantly reduce prolificacy and piglet birth weight, respectively (Honeyman 1997).

5.3 Gestation management

The goals of management for gestating sows include ensuring comfort, reducing aggression, detection of any animals returning to estrus, diagnosing pregnancy, ensuring proper individual feed intake, and reducing risk factors that increase the percentage of stillborns (Hutu, 2005).

5.3.1 Hoop barn designs for gestation units

The basic design requirements of a gestation unit are as follows:

1.	Resting area per sow	>2.25 m^2
2.	Service alley	0.9–1.25 m wide; rough concrete floor
3.	Feeding stalls	0.50 m wide × 2.15 m long; rough concrete

Basically then the gestation unit is divided into a:

- service alley;
- feeding area with stalls that also may serve as confinement units during the first 28 days of gestation;
- resting area group pen for housing sows for the remainder of gestation.

Figs. 5.8 through 5.13 illustrate various designs that work well for gestation units. Figs. 5.8 and 5.12 show a 64 sow unit that provides 6.2 m^3 per sow. The feeding platform is 40 cm above the floor of the resting area. Sows access it via four 10 cm steps. The feeding platform uses 29% of the total floor area of the barn. It consists of two rows of 16 feeding stalls each with a feeding alley in front of each row of stalls and an access alley between the two rows for sows. The feeding area is separated from the resting area by two gates (item 5 in Fig. 5.8), which are opened to allow one pen of 32 sows at a time to enter the feeding area. Feeding stalls may also be used for pregnancy detection, administration of injections, etc. The

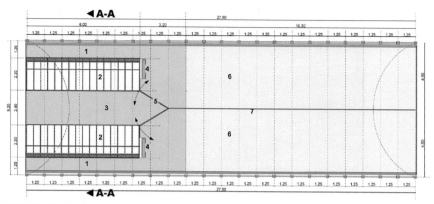

FIGURE 5.8A Floor plan for a 64 sow gestation unit.

1. service and animal movement alleys; 2. 32 individual stalls, two rows × 16 stalls; 3. central alley, for access to feeding stalls; 4. water fountains; 5. gates; 6. deep litter bedded area, 32 gestating sows per pen; 7. dividing wall in-between pens.

Redrawn after Harmon, J. D., Honeyman, M. S., Kliebenstein, J. B., Richard, T., Zulovich, J. M., 2004. Hoop Barns for Gestating Swine. Rev. Ed. AED44. MidWest Plan Service. Iowa State Univ., Ames, IA. p. 20.

feeding or service alleys at the front side of the feeding stalls may also be used for presentation of boars during heat detection for any sows returning to estrus.

Freeze proof water fountains are located at the end of each feeding stall row such that sows have ad libitum access from each of the two resting areas. There are then two resting area pens running longitudinally, which each houses 32 sows.

The design illustrated in Figs. 5.9 and 5.10 allows sows individual access to feeding stalls and also allows for dividing the resting area into 2 to 6 group pens depending on needs. The feeding platform is again 0.4 m above the resting area and has a rough concrete floor. There is a service alley 0.9 m wide that runs the length of the barn along one side in front of the feeding stalls. Again, with sows confined in the feeding stalls, estrus detection may be performed by moving boars down the service alley.

The resting area with deep bedding may be divided by gates oriented across the area. Such divisions are flexible and will depend on the sow group sizes. Access to feeding stalls for each subgroup will be limited to the stalls that face that section of the pen. All 55 sows will have access to an individual feeding stall at any time. During feeding time, sows may be locked into the stalls and animals with low body condition scores may be fed extra feed. Freeze proof water fountains may be installed on the wall opposite the feeding platform in each subdivision with a minimum of one fountain for every 15 sows. Water may also be provided by nipple waterers in the feeding stalls if sows are to be confined in them during breeding and early gestation (Figs. 5.10 and 5.13).

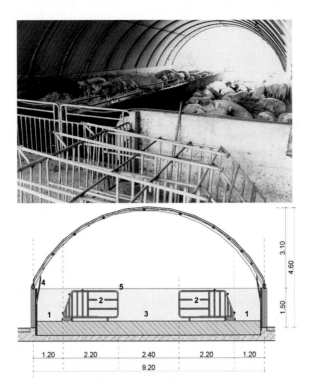

FIGURE 5.8B Layout and cross section of a barn for 64 pregnant sows with 2 rows of 16 feeding stalls each.

Upper: Pictures of deep litter bedded area, view from the feeding area. Lower: 1. service and animal movement alleys; 2. 32 individual stalls in two rows of 16 stalls; 3. central alley for access to feeding stalls; 4. air deflection panel; 5. feeding area end wall.

Credit: Upper Photo, Payne, H., 2004. Deep-litter housing systems an Australian perspective, Hoop Barns & Bedded Systems for Livestock Production Conference, 14 September 2004, Ames, Iowa, Australia. With permission from Honeyman, ISU Hoop Conference. Hutu, 2007, drawing adaptation after photo (lower).

Another design is illustrated in Fig. 5.11. This design incorporates an electronic sow feeder and will house 65 animals. Electronic feeders can control the daily ration each sow receives and therefore are an excellent way to feed for necessary body condition adjustments. The feeding platform is again 0.4 m above the floor level of the resting area. The size of the feeding area is merely large enough to accommodate the electronic feeder plus a small access area. The remainder of the barn is then free for a deep bedded resting area.

5.3.2 Sow group management

Before moving a group of sows into the unit, feeding and watering equipment should be checked over and necessary maintenance performed. Sows should be moved in

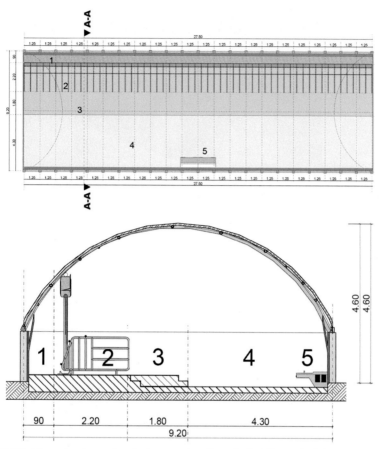

FIGURE 5.9 Floor plan (upper) and cross section (lower) of a 55 sow individual feeding stall barn.

1. Service and animal movement alley; 2. 55 individual feeding stalls; 3. optional access platform for feed stall area; 4. deep litter bedded area; 5. water fountain.

Redrawn after Harmon, J. D., Honeyman, M. S., Kliebenstein, J. B., Richard, T., Zulovich, J. M., 2004.
Hoop Barns for Gestating Swine. Rev. Ed. AED44. MidWest Plan Service. Iowa State Univ., Ames, IA.
p. 20 (upper). Draw by Hutu, 2007, Swine Experimental Units, Timişoara, Romania (lower).

groups following all-in-all-out management principles. Sows may be grouped by parity, weight, or weaning date. With the barn designs illustrated in Figs. 5.9 and 5.10, multiple groups of sows may be maintained in the same structure by subdividing the feeding and resting areas. Although this design is more costly, primarily due to the added number of feeding stalls that must be provided, it offers increased functional flexibility through the ability to group sows by body condition or size, etc. Grouping can occur at weaning, or at the time, sows are moved from breeding to gestation units. New sows to be added to existing groups should be placed within

FIGURE 5.10 Gestation barn with feeding stalls on one side.

The bedded area can be divided into pens for different groups of sows or for temporary storage of bedding (i.e., straw bales).

Photo: Lammers, P., Honeyman, M. S., 2004. Effects of bedded gestation housing on litter size and culling: preliminary results, Hoop Barns & Bedded Systems for Livestock Production Conference, 14 September 2004, Ames, Iowa, with permission from Honeyman, ISU Hoop Conference.

their own subgroups a few days before being introduced into the larger group to minimize the stresses from socialization. If group management during gestation is not performed well, decreases in prolificacy will be likely (Karlen et al. 2005).

Several management techniques can be used to counteract aggression when grouping sows:

- Maintaining sows in gestation stalls until the fifth week[8] of gestation before grouping
- Providing plenty of extra bedding for new groups (Durrel et al. 1997)
- Introducing a vasectomized boar into the sow pen for 3–5 days after regrouping[9]

To maintain a stable social hierarchy within sow groups once they are formed, the following management practices are helpful:

- Group sows by parity
- Introduce gilts as subgroups into larger groups and maintain static groups as much as possible with no resorting (Turner et al., 2003, 2004)

---

[8] Beginning about the fifth week of gestation, high levels of progesterone and estrogen mitigate aggressive behavior in sows just as they do in rodents.

[9] Presence of a boar in new groups lessens sows' aggressiveness and increases reproductive performance (higher embryo survival during implantation). The precise mechanisms for this effect are not well understood, but the presence of a boar (visual stimulation, direct contact, or pheromones) seems to redirect aggressiveness toward the boar itself or perhaps pheromones tend to quiet the sows. Boar androstenone does appear to sedate sows (McGlone and Morrow 1988).

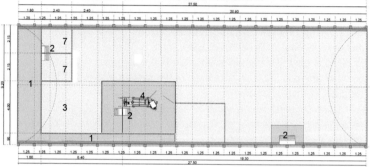

FIGURE 5.11 Photograph (upper) and floor plan (lower) of a 65 sow gestation barn with an electronic feeder.

1. Service and animal movement alley; 2. water fountains; 3. access area to electronic feeder; 4. electronic feeder on concrete platform 0.40 m above resting area floor; 5. gilt pens; 6. deep litter bedding area; 7. isolation pens.

Redrawn after Honeyman, M. S., Harmon, J., Lay, D., Richard, T., 1997. Gestating sows in deep-bedded hoop structures ASL-R1496 Management/Economics Iowa State, In: 1997 ISU Swine Research Report, AS-638, Dept, of Animal Science, Iowa State University, Ames, Iowa, USA; photo with permission from Honeyman, ISU Hoop Conference.

- Provide individual feeding stalls with continual access—these can serve as a refuge for sows subject to aggression at times outside of feeding
- Provide extra feed for thin sows
- Introduce gilts into sow groups after weaning of the sows or at the time of farrowing box removal and the beginning of group nursing
- Use corner pens or refuges to permit sows low on the hierarchy to have a place of escape if necessary (Wiegand et al. 1994)

These management practices will result in the sow herd experiencing fewer mobility issues, having better immune function and better body condition than would be the case otherwise.

FIGURE 5.12 Sows accessing a feeding area at one end of a gestation unit.

Resting area (upper and in back of picture) and feeding area (lower and in front of picture) with two rows of 16 feeding stalls each—also see Fig. 5.8. The resting area is divided into two pens of 32 sows each. One group is fed and when those sows have completed their feeding, the other group is fed. The advantage of this system is that only half as many stalls are required as there are sows.

Photo: Payne, H., 2004. Deep-litter housing systems an Australian perspective, Hoop Barns & Bedded Systems for Livestock Production Conference, 14 September 2004, Ames, Iowa, Australia. With permission from Honeyman, ISU Hoop Conference.

FIGURE 5.13 Gestation barn with rest area divided into multiple pens.

Feeding area and the bedded area divided into more pens for different groups of sows.

Photo: © State of Western Australia, Department of Primary Industries and Regional Development.

5.3.3 Management of barn environment

With the sows housed in a hoop structure with deep bedding, the animals will maintain their microclimate through proper interaction with the bedding and other barn areas as indicated in previous chapters. The use of feeding stalls is critical to proper management as it allows sows to avoid conflict during feeding and allows for confinement of sows for other purposes as well. Electronic feeders are quite expensive but do allow for precise control of feed amounts for each individual with minimum labor input.

Supplemental ventilation or water mist systems may be desirable in gestation units to keep inside temperatures down during hot spells of summer weather, therefore mitigating the adverse effects of high temperature on sow performance (Honeyman and Kent 2000).

Animal performance can also be enhanced by providing artificial lighting to maintain 14−16 h of light per day at 55 lumen/m^2.

Bedding usage is dependent on season (Table 3.3) with winter use higher than that of summer. In general, sows will need 1 kg bedding/sow/day resulting in about 110 kg per sow for the entire gestation. The manure pack is therefore quite dry, and water may need to be added to allow for proper fermentation during composting (Honeyman et al. 1997).

5.3.4 Sow space requirements and management

The gestation period should be divided into two phases:

- During the first 28 days, sows are housed in the feeding stalls or gestation crates in a separate breeding structure.
- Between days 29 and 110, sows are housed in a deep bedded group pen; group housing as such tends to enhance estrus in those sows that return to heat (Honeyman and Kent 1996).

To minimize embryonic death, abortions, and stress, it is important to have individual feeding stalls and adequate resting area space (Honeyman and Kent 1996). Group pens should be large enough that sows can maintain a 1.8 m social separation, which will minimize aggression due to social hierarchical interactions. Therefore, the resting area pen width should be a minimum of 4.6 m and sow population per pen maintained at 8−12. This will allow enough room for individual avoidance of aggression from "boss" animals. Sows should have a minimum of 2.25−2.50 m^2 resting area each, but this amount is dependent on the group size (Honeyman et al. 1997).

For large groups of 30 or more, 2.6 m is desired; for groups of 15−29, a greater area of 3.25 m^2 per sow is desirable; and for small groups of 10 or less, 3.7 m^2 per sow is recommended.

Thin sows may be isolated from the remainder of the group either in pens specifically for that purpose (Fig. 5.9) or in the feeding stalls (Fig. 5.13).

5.3.5 Vaccination and other veterinary procedures

Required vaccinations will depend on the health history of the farm. Typically, however, vaccinations for atrophic rhinitis, leptospirosis, parvo virus, *Clostridium* species, and *Escherichia coli* need to be given during each reproductive cycle. Leptospirosis vaccinations for naïve animals (gilts) require two doses spaced 14−21 days apart and administered a minimum of 6 weeks before breeding. Previously, vaccinated sows require a single dose at weaning.

Protection for piglets from pneumonia and diarrhea can be achieved by vaccinating sows before farrowing for *Pasteurella*, *Clostridium*, and *E. coli*. These antigens may be administered simultaneously and should be given to gilts at 6 weeks and again at 3 weeks before farrowing; sows at second parity or higher require a single dose of vaccine 3 weeks before farrowing if they were previously vaccinated at parity 1 (Moga 2001).

Pregnancy detection is typically performed between days 30 and 75 of gestation. The first indirect indicator of pregnancy is nonreturn to estrus. Boars can be paraded in the service alleys after feeding to facilitate detection of return heats. Sows not returning to estrus should be checked with ultrasound equipment to confirm pregnancy diagnosis. This equipment may be of a variety of types from simple inexpensive single crystal detectors (A mode) to rather costly Doppler or real-time (B mode) ultrasound units. Normally, two ultrasound examinations are performed—one at about day 36 and a second at about day 56. The procedure can easily be performed with sows locked into feeding stalls or some other confined areas.

5.3.6 Nutrition of gestating sows

The nutrient requirements of gestating sows are dependent on the sow's weight at mating time, the amount of weight gain required during gestation to reach proper body condition at farrowing, and the expected litter size.

For 140−205 kg sows requiring 40−65 kg weight gain and carrying 12−15 piglets, daily intake must consist of 6700−6900 kcal ME, 110−140 g SID protein (140−205 g CP), which includes 6.3−10.6 g SID lysine (7.7−12.4 g total lysine), 13 g Ca, 5.5 g available P, 8400 IU vitamin A, 1680 IU vitamin D, 92 IU vitamin E, and 2 g linoleic acid for the first 90 days of gestation. Between gestation days 90 and 110, total feed intake per day should be increased by a minimum of 0.4 kg, which will effectively increase the levels of all the above nutrients.

Feed amounts are generally limited to the amount needed to meet the animal's required energy intake. Feed delivery can be via a variety of methods including manual distribution, conveyor systems with drop boxes, automatic trickle systems, or electronic feeders. Table 5.5 outlines a feeding schedule for gestating sows and gilts. Note that feeding levels vary with the stage of gestation (Young et al., 2003).

A typical goal is to have gilts (137−150 kg at breeding) and small sows (165 kg at breeding) to gain a total of 60−65 kg during gestation (includes gains in tissue mass of the mother as well as the weight of the fetal tissues and fluids) and to reach

Table 5.5 Sow feeding levels (kg feed[a]/day) between days 0 and 115 of gestation for various body weights, P2 fat depths (FD), and body condition scores (BCS).

Day of gestation	FD (mm)	BCS	Sow	Gilts
d 0–5			2.15	1.90
d 6–35	<15 mm	1	3.85–4.80	3.6–4.55
	15–17.9 mm	2	2.90	2.65
	18–20 mm	3	2.40	2.15
	20.1–23 mm	4	2.15	1.95
	>23 mm	5	2.15	1.95
d 36–100	<18 mm	<3	2.40	2.15
	≥18 mm	>3	2.15	1.95
d 101–115		+1 kg in addition to d 100 feed level.		

[a] *Adapted after Young, M. G., Aherne, F. X., Main, R. G., Tokach, M. D., Goodband, R. D., Nelssen, J. L., Dritz, S. S., 2003. Comparison of three methods of feeding sows in gestation and the subsequent effects on lactation performance, Kansas Agric. Exp. Station Res. Reports 0, 10. https://doi.org/10. 4148/2378-5977.6823 and Luca, I., 2000. Curs de alimentatia animalelor, Ed. Marineasa, Timişoara, for mixed feed with 3100 kcal ME/kg.*

25 mm back fat by farrowing. During the past 2 weeks of gestation, feed amounts may be increased by 1–1.5 kg per day to a maximum of 4.5 kg per day.

Overfeeding may result in decreased appetite during lactation. Heavy sows (>200 kg at breeding) should gain no more than 40–45 kg during gestation and should be carrying no less than 15 mm back fat at farrowing. Honeyman and Kent (2001) suggest an energy dense feed consisting of 3300 kcal ME/kg, 12.2% DP, and 0.56% lysine be fed at a rate of 1.8 kg per day for the first two-thirds of gestation and 2.7 kg per day for the remaining third.

Additionally, in winter, to maintain body temperature, sows will need 5%–25% more feed per day (Holden et al. 1996; Lammers et al. 2006). Thinner sows will require the higher percentages, and fatter sows will require the lower increases during cold weather.

Sows should be condition scored at weaning and assigned gestation feeding levels based on that score. They should be rescored on day 60 of gestation and feeding levels adjusted if needed to attain a final score of 3.0–3.5 at the next farrowing.

Gilts that have been properly developed who have attained a weight between 135 and 150 kg at breeding should receive 2.0–2.25 kg of feed/day (3300 kcal ME/kg) during gestation (Crainiceanu and Matiuti, 2006).

In Romania typical feed (3050–3150 kcal ME/kg), ingredients include the following:

- Cereal grains (corn, barley, oats) at 60%–70% of the ration
- Protein feeds (bran, oil meals, peas) at 20%–25% of the ration

- Animal protein sources (meat and bone meal, blood meal) at 2%–3% of the ration
- Mineral feeds: salt 0.5%; feed lime 1%–1.5%; phosphates 0.5%–1%; trace mineral and vitamin premix 1% (Luca 2000).

Table 5.6 outlines a variety of possible gestation rations (Selk, 2006). Some authors suggest that in hoop barns, the consumption of bedding materials during gestation enhances appetite during the subsequent lactation perhaps through enlargement of the digestive tract volume (Matte et al. 1994).

5.3.7 Expected production performance

Most research indicates no significant effect on reproductive performance owing to housing sows in cold hoop structures.

Farrowing rate has been reported to average near 75% for hoop housed sows (Honeyman and Kent 2000; Lammers et al. 2006; Leaflet 2006), and conception and farrowing rates for hybrid gilt lines have approached 95% in hoop housing systems (Honeyman et al. 1997). Ewan et al. (1996) suggest that bedding consumption during gestation actually increases litter size.

Further data with ¼ Hampshire × ¼ Landrace × ½ Yorkshire sows mated to Duroc boars indicated less than 20% of sows returning to estrus following the first breeding and subsequent farrowing rates of greater than 80% (Honeyman et al. 2002).

5.4 Farrowing management

The goal of management practices in farrowing barns is to allow sows to exhibit their natural biological and mothering instincts. These behaviors are manifested in the following order: voluntary isolation from other sows just before parturition; building of a nest; farrowing proper; nesting for nursing and rest; leaving the nest to join other sows; and weaning. Cold barns are not ideal for farrowing, but opportunities do exist to utilize them effectively as indicated by the following observations:

- Hoop structures may be partially insulated and zone heated to maintain the minimum 10°C barn temperature.
- Farrowing of sows in temperature-controlled housing and nursing in deep litter bedding is a basic characteristic of the Ljungstrom version of the Swedish system where sows farrow in conventional barns, and at 14–20 days of age, piglets are grouped and nursing is moved to common pens with deep litter bedding.

Table 5.6 Various gestation rations and their nutrient levels.

Ingredient	Various energy and protein proportions							
	R1	R2	R3	R4	R5	R6	R7	R8
Corn, %	81.40		69.10					
Sorghum, %		77.40			76.90	68.60		
Wheat, %				81.00			85.40	80.40
Wheat bran, %			15.00			15.00		
Alfalfa meal, %		5.00			5.00			
Soybean meal, %	16.00	15.00	13.00	16.40	15.50	13.50	12.00	12.00
Limestone, %	0.90	0.90	1.00	1.00	0.90	1.00	0.90	0.90
Dicalcium phosphate, %	0.80	0.80	1.00	1.00	0.80	1.00	0.80	0.80
Salt, %	0.50	0.50	0.50	0.50	0.50	0.50	0.50	0.50
Minerals and vitamins, %	0.25	0.25	0.25	0.30	0.30	0.30	0.25	0.25
Lysine, %	0.10	0.10	0.10	0.10	0.10	0.10	0.10	0.10
Energy, ME kcal/kg	**3295**	**3213**	**3127**	**3406**	**3318**	**3220**	**3283**	**3200**
Protein, %	14.45	14.45	14.26	15.51	15.51	15.23	15.15	15.42
Lysine, %	0.66	0.65	0.61	0.64	0.63	0.60	0.65	0.65
Tryptophan %	0.14	0.13	0.11	0.14	0.14	0.12	0.18	0.18
Threonine, %	0.43	0.43	0.41	0.44	0.44	0.43	0.44	0.45
Met. + cystine%	0.43	0.42	0.47	0.39	0.39	0.44	0.54	0.53
Total calcium, %	0.65	0.70	0.74	0.65	0.71	0.74	0.65	0.70
Total phosphorus, %	0.50	0.50	0.64	0.51	0.51	0.64	0.52	0.52
Fermentable fiber, %	9.64	10.49	10.13	6.71	7.72	7.67	7.73	8.82

After NRC 2012 requirements, calculation for 114 days of gestation, 7095 kcal/day ME intake for sow weighing 165 kg at breeding and 220 kg at farrowing; anticipated litter size of 12 with 1.30 kg average birth weight.

5.4.1 Hoop barn design for farrowing units

Although farrowing is difficult to manage in tarp (insulated version)-covered hoop barns, if such a structure is to be built, the following specifications should be followed:

Farrowing box size:	Minimum of 1.8 × 2.4 m
Air volume per sow:	Minimum of 21 m^3
Access for feed:	Minimum of 1.8 m wide, rough concrete
Feed area:	25% of useful barn area, 0.4 m above level of bedded area

Such barns should be compartmentalized such that there is adequate provision for sows to nest, have easy access to feed and water, and for piglets to stay in the nest (up to a certain age). Compartmentalizing will also facilitate sorting and grouping when farrowing boxes are dismantled.

Figs. 5.14 and 5.15 illustrated two designs for farrowing units to be used in warm or relatively warm climates. The farrowing unit is typically divided into two longitudinal regions each of which in turn is divided into three areas as described below.

Feeding area: A concrete platform at 0.40 m above the level of the remainder of the shelter located either at the south end (Fig. 5.14) or in the center (Fig. 5.15). Sows have free access to feed and water at these sites. It is typical to establish a creep area for piglets upon reaching 7−10 days of age on the feeding platform or in a

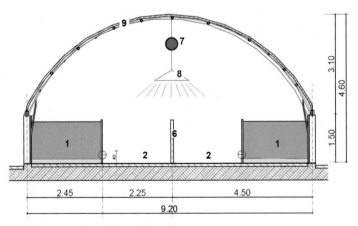

FIGURE 5.14A Cross section of farrowing unit.

1. Farrowing boxes; 2. alley to feed and water; 7. radiant tube; 8. lamps; 9. temperature insulating tarp.

Adapted after Lammers, P., Honeyman, M. S., Harmon, J., 2004a, Alternative systems for farrowing in cold weather. Agricultural Engineers Digest, AED 47, September, 2004, Available at: http://www.nlfabric.com/assets/documents/alternative-farrowing-systems.pdf.

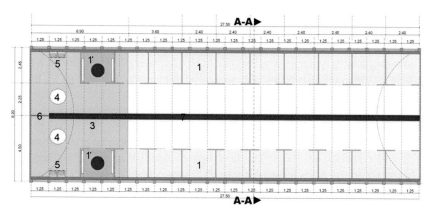

FIGURE 5.14B Floor plan of a farrowing unit 1.

Farrowing boxes (9–10 sows per compartment and 18–20 sows per barn); 1'. boxes convertible to creep feed area; 2. access alley to feed and water; 3. feeding area with concrete floor; 4. feeders; 5. water fountains; 6. divider; 7. radiant tube.

Adapted after Lammers, P., Honeyman, M. S., Harmon, J., 2004a, Alternative systems for farrowing in cold weather. Agricultural Engineers Digest, AED 47, September, 2004, Available at: http://www.nlfabric.com/ assets/documents/alternative-farrowing-systems.pdf.

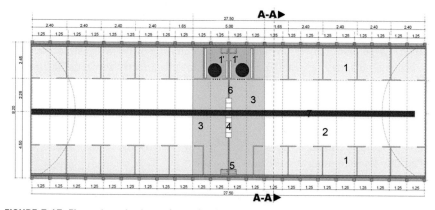

FIGURE 5.15 Floor plan of a farrowing unit with a centrally located feeding area.

1. Farrowing boxes (9–10 sows per compartment and 18–20 sows per barn); 1'. creep area; 2. alley to feed and water; 3. feeding area; 4. feeders; 5. water fountains; 6. divider; 7. radiant heater.

Redrawn after Payne, H., 2004. Deep-litter housing systems an Australian perspective, Hoop Barns & Bedded Systems for Livestock Production Conference, 14 September 2004, Ames, Iowa, with permission from Honeyman, ISU Hoop Conference.

farrowing box not habituated by a sow and litter. The creep area will contain 2 or three water nipples, feeders with prestarter feed, and milk replacer dispensers. Sows are kept from entering by steel gratings.

Farrowing area: It typically contains one row of farrowing boxes along each side-wall of the structure. Boxes are 1.8–2.4 m in dimension and 1.2 m in height (Hilbrands et. al. 2004). On the side facing the alleyway, each box should have an opening for the sow to exit and enter, measuring 50 × 70 cm and fitted with a roller on the bottom edge, which serves to protect the sow's udder and keep small piglets confined in the box (Figs. 5.14 and 5.15) during the first week or so after farrowing.

The roller should be 10–15 cm in diameter, and its upper side should be 35–40 cm above the floor level (Algers 1991; Braun and Algers 1993). As soon as piglets are large enough to climb over them, front gates of the boxes may be removed.

Alleyway allowing access to feeding area: This area is distinct only during the time the farrowing boxes are in place. It is typically a 2.25–2.35 m wide corridor that allows sows ready access to the feeding area. After farrowing boxes are removed, the entire region where the boxes and alleyway were now becomes the resting area and is deep bedded for a common nursing pen. The floor is usually rough concrete with a 1.5% slope from south to north but no slope side-to-side.

The tarp cover for farrowing units is usually the air insulated type. Normally, no cooling or heating equipment is installed in hoop barns as the heat from the fermentation of the bedding pack, and the insulating effect of the bedding in the nest provides adequate warmth for piglets (Honeyman and Kent 1996).

In certain climates, however, heating or cooling equipment may have to be installed. Large units may be installed on the centerline of the hoop structure, and zone heating units may be installed along the sidewalls directly above the farrowing boxes. It is advisable to install light fixtures along the centerline of the barn so that farrowing boxes receive only indirect light.

Figs. 5.16 through 5.18 (middle) and Fig. 5.19 (upper) illustrate a different approach for compartmentalizing the farrowing unit, which is more suitable for production in colder climates. Again, there are two farrowing compartments each divided into three regions as follows:

Feeding area: Same specifications as in the previous design and located either at one end or the center of the barn. The creep area is provided with a 250 W heat lamp in this case.

Farrowing area: This will again consist of two rows of farrowing boxes except that now they are on the centerline of the structure rather than along the sidewalls and are covered with a second hoop that is heated with a 150,000 BTU (British thermic units)/hour radiant tube. This is a "hoop-within-a-hoop" design (Fig. 5.16).

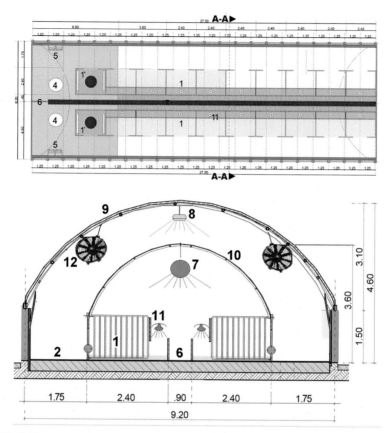

FIGURE 5.16 Floor plan and cross section of a farrowing unit designed for colder weather use.

1. Farrowing boxes with heated creep area; 2. alley to feed and water; 3. feeding area; 4. feeders; 5. water fountains; 6. divider (upper) or service alley for heated creeps (lower); 7. radiant tube; 8. lights; 9. temperature insulated tarp; 10. inner hoop; 11. creep area warmed with heat lamps; 12. sow cooling fans.

Redrawn after Lammers, P. J., Honeyman, M. S., 2005, Farrow-to-Finish in a Hoop Barn: A Demonstration, Animal Industry Report: AS 651, ASL R2029. Available at: https://lib.dr.iastate.edu/ans_air/vol651/iss1/29. With permission from Honeyman, ISU Hoop Conference preview figure.

Alleyway for feed access is 1.6 m wide and is located along the sidewalls of the structure. Again, it is only functional while the farrowing boxes are in place. Two to four 75-cm fans placed above the alleyway will allow for good summer ventilation.

The outer tarp cover may be a simple single layer type or a double air insulated type. The tarp covering the farrowing box area is typically a single layer transparent type, which will maintain a higher temperature in the center section of the barn during cold seasons.

FIGURE 5.17 Farrowing boxes located within an inner hoop for added heating capability.

The farrowing boxes and inner hoop (upper). Sow and litter in farrowing boxes within an inner hoop. Farrowing box with guard rail for piglet refuge (lower).

After Honeyman M.S., Lammers P., 2004. Deep bedded systems for sows and piglets, In: Alternative Swine Housing Systems: An International Scientific Symposium, 15 September 2004. Ensminger Room, Kildee Hall, Iowa State University Ames, Iowa. Checkoff production systems committee (www.pfi.iastate.edu), 2004. With permission from Honeyman, ISU Hoop Conference.

5.4.2 Management of the farrowing unit

Depending on the available existing facilities, farrowing can be in insulated conventional barns (as per Ljungstrom[10]) or in cold barns. When using hoop structures, the following practices are recommended. The farrowing group of sows will be moved

---

[10] Ljungstrom version (from Gunnar Ljungstrom, a Swedish farmer who invented the system, Halverson, 1994): Farrowing takes place in insulated barns. Sows are introduced to the farrowing boxes 2—3 days before farrowing and remain in the farrowing unit for 14—20 days. Farrowing units on small farms typically contain two regions of 20 boxes each and are managed very intensively with 16—18 production cycles per year. Sows and piglets are moved to deep bedded group nursing pens in hoop barns at 14—20 days. Movement occurs over the span of a day with gilts, and their litters moved first, followed by small and medium sized sows when the gilts have become acclimated. Finally, large sows and litters are moved last. This helps mitigate the aggressive behavior of the larger older sows as they arrive last and must concentrate on caring for their litters.

FIGURE 5.18 Group farrowing and group nursing.

Access alley along boxes before parturition and first 10 days of lactation (upper left). Farrowing box with a roller in the entryway (upper right). Group lactation with sows and piglets outside farrowing boxes (lower left). Creep area with two nipple drinkers and two feeders—one for grain feed and one for milk replacer (lower right).

Photo: Payne, H. G, Nicholls, N., Davis, M., 2004. Group farrowing in a deep-litter eco-shelter, Alternative Swine Housing Systems: An International Scientific Symposium, September 15, 2004 Ensminger Room, Kildee Hall, Iowa State University Ames, Iowa, Australia. With permission from Honeyman, ISU Hoop Conference.

in following cleaning of the unit and cleaning of the sows. Farrowing groups should be homogenous grouped either by size or by parity. Groups should enter the farrowing unit 2–5 days before the expected farrowing date—usually at day 110 of gestation. Group size should be 8–12 sows.[11] There should be at least one farrowing box available for each sow. Sows may be culled from the group or added to the group at weaning and 2 weeks postbreeding, respectively (Honeyman and Kent 1996).

Sows are free to build nests in the farrowing box of their choosing, but not all do so. Ideally, 98% of sows would build nests in the boxes, but in practice, however, without assistance about 90% (Baxter 1991), 88% (Fisher 1990), or 83% (Nash 1994) have been the reported to do so in research trials.

---

[11] Rossiter and Honeyman 1997 suggest a 9.2 × 22.0 m space for each group of 30 sows, therefore allowing 0.9–1.1 m^2 for each finishing hog from the litters. This suggestion assumes that 80% of sows (breeding 30 will result in 24 weaning litters) will wean an average of 8–9 piglets each for a total of 190–215 piglets per farrowing unit.

Sows can be encouraged to nest in the boxes by the following strategies:

- Discourage farrowing outside boxes by keeping floors wet and well lit (Houwers et al. 1992).
- Lock sows in farrowing boxes before parturition (Houwers et al. 1993).
- Facilitate farrowing at any location in the unit by deep bedding the entire floor area; this is an adaptation from the Thorstensson[12] version (Algers 1991).

During farrowing, assistance is typically not necessary in this system. Crushing of piglets can be minimized by the installation of guard rails along the entire perimeter of the farrowing box to give piglets an escape route when the sow lies down next to the side of the box, which is typically where she will lie. The guard rails should be about 25 cm above the floor level. Fostering of piglets may be done during the first 2 days to equalize litter sizes. Fostered piglets should be put with sows having the same age litter (Honeyman and Kent 2001) and is not a source of problems as long as piglets are homogenous with regard to weight and age (Algers 1991). When piglets are 7–10 days old, they will begin to climb over the rollers in the box openings and comingle with one another. Front farrowing box panels and rollers can be removed and by 10–14 days. The farrowing boxes themselves should be removed to allow for complete comingling of litters (Payne et al. 2004) sometime during the first 2–5 weeks. After comingling, the sows and litters will inhabit the entire resting area. In this system, there is considerable cross-nursing between all the litters and sows (Honeyman and Kent 1996).

Weaning should take place at 5–6 weeks of age. In the European Union and in Romania, animal regulations require that weaning not take place before 28 days of age. Weaning is accomplished by removing the sows from the farrowing area to either the breeding unit or culling to market. Weaning stress on the piglets is minimized by removing sows and leaving the piglets in the farrowing area to which they are accustomed rather than removing them as well. Milk production in the sows may be reduced before weaning by gradually restricting their feed intake for a few days and by withholding feed on weaning day. If high milk flow is a problem in spite of feed reduction, other strategies that will assist with cessation of lactation include the following:

- Movement to the estrus detection and breeding unit
- Administration of vaccines

5.4.3 Environmental control of farrowing unit

When farrowing in a cold environment, barns must be insulated. The farrowing area should be set up with farrowing boxes installed so that sows can be moved in 2–5 days before farrowing. The presence of the boxes reduces aggression as it

---

[12] Named for the Swedish farmer Goran Thorstensson.

decreases visual contact between animals (Arey et al. 1992). At any rate, sows tend to seek isolation shortly before farrowing, and the boxes allow this to happen (Jensen 1986). As described earlier, the farrowing area will include the boxes and an access alley for sows to get to the feeding area. The boxes should be bedded with 15–30 cm of straw. The alleyway is left unbedded to discourage sows from farrowing there. In addition, during warm seasons, the unbedded concrete floor offers a way for sows to cool themselves through conductive losses to the cooler concrete as they lie on it. Feeding areas can be equipped with self-feeders as sows should be receiving essentially ad lib feed at this time.

The temperature should ideally be in the range of 15–20 °C for the sows, and air flow should be around 25–200 m^3 per hour per animal depending on the season.

Heat may be provided in localized zones by heat lamps located in the farrowing boxes or may be provided for the entire area using radiant tubes located under the tarp of the inner hoop (Figs. 5.16 and 5.17).

The hoop may be insulated through using the double-layer tarps with trapped air between the layers. Temperature regulation for the design illustrated in Fig. 5.16 is determined by the season (Lammers et al., 2004, 2006):

- During winter, the inner hoop is placed over the farrowing boxes and heat is provided via a 150,000 BTU radiant tube supplemented with heat lamps in the farrowing boxes. The creep area is also locally heated with a 250 W heat lamp. The energy required for the 5 weeks of one farrowing cycle is 3.33 m^3 of liquefied propane (LP gas) and 5020 kWh of electricity or 0.166 m^3 of LP and 251 kWh per sow and litter.
- During summer, the inner hoop over the farrowing boxes is not needed nor is any supplemental general or localized heating except for the creep area where a 250-W heat lamp may be installed. Two fans are used in the alleyways to facilitate cooling the sows. Each fan should deliver 11,400 m^3 air/h. Energy consumption for this arrangement will be 690 kWh (404 for the creep area and 286 for the fans) for one farrowing cycle or 34.5 kWh per sow and litter.

Lighting should be provided for 18 h per day at a minimum intensity of 100–150 lumen/m^2.

The bedding material should be straw replenished daily. The amount of straw required is typically about 900 kg for a 10 or 11 box farrowing room up until boxes are removed. An additional 600 kg will typically be used for the remainder of the nursing period until weaning.

5.4.4 Sows and piglets

Housing of sows and piglets will initially be in the 1.8 × 2.4 m farrowing boxes and later, after removal of the boxes, in the common nursing area, which typically will accommodate 8–12 sows and litters. Although farrowing boxes are in place, sows are free to leave them and access feed and water in the feeding area. Early on, piglets will be confined to the box but at about a week of age will begin to climb out of it.

Once boxes are removed, each sow and litter has access to about $8\,m^2$ of deep bedded space (Halverson and Honeyman, 1997) in the common nursing area in addition to being able to access the feeding area. A creep pen should be set up in the feeding area for piglets (Fig. 5.19) to allow access to prestarter feed and milk replacer in addition to water nipples. Sows can obtain their feed from self-feeders with one feeding hole allotted for each sow.

5.4.5 Health management

At a minimum of 2 weeks before farrowing, sows should be vaccinated for *E. coli* and *Clostridium* (Lammers et al. 2006). On gestation day 110, when moved into the farrowing area, sows should be dewormed using an anthelminthic such as ivermectin. Two weeks before weaning, sows should receive vaccinations for reproductive diseases such as leptospirosis (Moga-Mânzat, 2001).

Typical illnesses that affect farrowing sows (such as metritis, mastitis, or agalactia) are uncommon in this system. This is likely the result of the increased freedom of movement, the availability of bedding for nesting and consumption, and the overall lower stress levels associated with this system (Honeyman and Kent 1997).

Piglets may have their needle teeth clipped between days 1 and 2 following birth to prevent injury to sow's nipples, although recent research would indicate it is best not to perform this procedure (Gallois et al. 2005) because high incidence of tooth and gum damage to piglets with no definitive benefit for sows. Teeth should not be clipped before 6 h after birth to allow piglets time to consume adequate colostrum without interference. Also within the first 2 days of life, piglets should be administered iron dextran shots to prevent anemia. Piglets should receive rhinitis and erysipelas vaccinations at about 3 weeks of age (Moga-Mânzat, 2001). Piglets should be treated with dewormer at weaning as well.

5.4.6 Nutrition of lactating sows and piglets

Feed will be offered ad libitum to lactating sows, and rations should be designed to optimize milk production and control body condition loss in the sows.

If temperature is maintained in the 18–20°C range for sows weighing 175 kg postfarrowing and nursing 11 piglets and losing about 15 kg of body weight over 30 days, daily nutrient intakes should be as follows: 18,700 kcal ME, 786 g total protein of which 56.5 g is lysine, 45 g Ca, 22.6 g P, 11,930 IU vitamin A, 4770 IU vitamin D_3, 262 IU vitamin E, and 6 g linoleic acid. For feeds with an energy density of 3300 kcal ME/kg, sows must consume about 5.9–6.6 kg per day to meet their maintenance and milk production requirements.

The feeding regime for sows is dependent on the stage of lactation and may be carried out as follows:

- A gruel consisting of 1.0 kg bran and 5.0 L water on farrowing day to serve as a laxative if there is a history of constipation within the sow herd.

FIGURE 5.19 Problems associated with farrowing boxes and group nursing.

Doubling up; straw management; sows calling piglets out; cross-suckling; dispersion in pen and large litters associated with high mortality (left to the right and top to bottom).

Credit: Payne, H. G, Nicholls, N., Davis, M., 2004. Group farrowing in a deep-litter eco-shelter, Alternative Swine Housing Systems: An International Scientific Symposium, September 15, 2004 Ensminger Room, Kildee Hall, Iowa State University Ames, Iowa, Australia, With permission from Honeyman, ISU Hoop Conference.

- A progressive increase in the amount of total feed per day from 2.5 kg up to 5 or 6 kg within the first 2−10 days postfarrowing; many swine nutritionists recommend rapid increases in feed immediately following farrowing. A recent field trial by PIC International found that sows whose feed intake was rapidly increased (3 kg increase/day) or who were allowed ad libitum intake immediately after farrowing experienced a reduction in percent of sows not bred by 7 days postweaning from 28% for the standard step-up feeding program to 16% for the aggressive or ad lib programs (Culbertson, 2014).
- Ad libitum feeding for the remainder of the nursing period up until 3 days before weaning.
- Maintain lactation intake during breeding week for improved reproductive performance.

In practice in Romania, a lactation diet may be formulated as follows:

- corn 40%−50%, barley and/or oats 20%−25%;
- bran 10%−15%, sunflower or soybean meals 8%−12%;
- alfalfa meal 5%−10% and meat and bone meal 2%−3%;
- salt 0.5%, limestone 1%−1.5%, phosphates 0.5%−1.0%, and vitamin trace mineral premix at 1.0% (Luca 2000).

Feed intake by sows will vary with temperature. During periods of high temperatures, consumption will drop and intervals between meal taking will increase. It is advisable to feed during the nighttime when temperatures are cooler (Gourdine et al. 2004; Renaudeau et al. 2005).

Once the farrowing boxes are removed, piglets will have access to the sows' feed and will begin to consume some of it. In addition, they will begin to consume creep feeds. By 28 days of age, piglets should have consumed a minimum of 0.6 kg of prestarter creep per head. This prestarter feed should be continued until days 7−14 following weaning (Payne et al. 2004) or until piglets have reached a cumulative consumption of 1.4 kg[13] of prestarter or starter. Creep feeding tends to improve uniformity of size among the piglets, makes the transition at weaning less stressful for piglets, improves overall growth rate, and decreases days to market.

5.4.7 Expected production performance

Farrowing in cold barns with deep bedding presents some unique management challenges that are not always effectively dealt with. Some of these include the following:

- two sows attempting to farrow in the same box;
- uneven spreading of bedding;

---

[13] Feed composition should consist of 3130 kcal ME/kg, 21.0% protein with 1.4% lysine (Honeyman and Kent 2001).

- unintentional fostering of piglets onto sows other than their dams;
- rejection of piglets by their mothers;
- increased piglet losses from crushing.

Most authors suggest that with properly managed hoop barn housing systems, farrowing performance is acceptable (Table 5.7).

The number of liveborn piglets, birth weight, and other parameters are essentially comparable with other systems (Honeyman et al. 1997; Honeyman and Kent 2000). Lammers et al. (2006) found that ¼ Hampshire, ¼ Landrace, and ½ Yorkshire hybrid sows inseminated with Duroc semen had an advantage of 0.8 more liveborn per litter and 0.4 more pigs weaned when managed in hoop structures as compared with conventional systems. Preweaning mortality in alternative systems is higher than in conventional ones by about 1%–2% (Halverson 1991a). However, when ambient temperature falls below −15 °C crushing deaths may approach 40% in alternative farrowing box systems. As most of these deaths occur during the first 2 weeks after farrowing, short-term supplemental heating is a realistic cost-effective solution to this problem (Rossiter and Honeyman 1997; Payne et al. 2004).

The longevity of sows is dependent on a number of factors—perhaps the most important being genetics (breed[14] or hybrid line). Age and body condition of gilts at initial mating also affects longevity. Type of housing and length of lactation are two additional factors. Lactation lengths of 28–35 days seem to be optimum for extending sow productivity as both shorter or longer lactations result in lower conception rates, longer weaning-to-estrus intervals, and increased culling rates (Svajgr et al. 1974).

Housing on deep bedded packs significantly reduces cull rate compared with sows raised on slatted floors (D'Allaire et al. 1989) likely as a result of better feet and leg health, decreased udder injuries (Ehlorsson et al. 2002), or decreases in digestive ailments (Christensen 1995). Inspection of the differences in reproductive performance between breeds or hybrid lines reveals that the most specialized breeds are the ones most prone to losses of performance under adverse conditions.

Native or rustic breeds seem to perform relatively better under the tougher conditions found in cold barn housing systems.

5.5 Grow-finish management

Management of the grow/finish stage of production in cold hoop structures requires careful attention to the biology of the animals and how that interplays with their environment.

---

[14] This is an important consideration in choosing sow genetics. Romanian native breeds such as the Mangalita and Bazna (or even Black Strei) have been shown to be extremely resistant to environmental stress and are also recently highly regarded in the pork industry for their highly marbled meat.

Table 5.7 Reproductive performance of sows housed in various gestation systems.

Parameter	Gestation crates[a]	Group pens[b]		Hoop barns[c]	
		Feeding stalls	Elect. feeder	Feeding stalls	Elect. feeder
Birthing and piglet performance					
No. born, unit	10.2	9.2	9.3	9.2	9.5
Stillborn, unit	0.84	1.07	1.00	0.58	1.17
Mummified, unit	0.14	0.21	0.14	0.27	0.11
Birth weight, kg	1.53	1.55	1.62	1.52	1.55
Number weaned, unit	9.18	7.54	8.11	7.54	7.60
Weaning weight, kg	5.17	5.21	5.12	4.94	5.35
Wean age, days	18.3	17.7	17.9	18.0	18.2
Average daily gain, g	199	207	196	190	209
Sow performance					
Prefarrow weight, kg	193.2	191.0	187.8	178.3	188.7
Back fat at farrowing, mm	35.1	34.3	33.3	33.0	33.3
Postwean weight, kg	163.7	160.1	161.0	158.8	161.0
Back fat at weaning, mm	33.3	31.5	30.7	31.8	31.5
Daily lactation feed intake, kg	4.4	3.9	4.3	4.6	4.4

[a] Sows in individual gestation crates provided with mechanical ventilation and partially slatted floor with flush gutters.
[b] Gilts in group pens with natural ventilation via curtain walls; partially slatted floor, no bedding, and feeding either in individual stalls or by electronic feeders.
[c] Gilts in hoop barns with natural ventilation; deep litter bedding and feeding in individual stalls or electronic feeders.
After Honeyman, M. S., Kent, D., 2000. Reproductive Performance of Young Sows from Various Gestation Housing Systems, 1999 Swine Research Reports, AS-642, Iowa State University, Ames, IA.

5.5.1 Hoop structure design for grow/finish units

Structures for grow/finish hogs are the simplest to erect as fewer components are required, and the management is less complex than for breeding, gestation, or farrowing operations. To optimize management and minimize labor requirements, the following structural parameters are suggested:

1. Available area/animal — $1.1\ m^2$
2. Feeding area — 25% of usable floor area, rough concrete, 0.4 m above grade of resting area
3. Resting area — 75% of usable floor area

A variety of barn types work for grow/finish units.

Hoop structure with feeding area located at one end. If farrowing boxes or feeding stalls are removed from farrowing or gestation units, respectively, they may be used for finishing hogs as well. (Figs. 5.20 through 5.24 illustrate some different designs.) Fig. 5.20 illustrates a design with the feeding area (item 2 in the figure) located at the south end of the structure. Such a barn may be operated as one large pen or may be divided longitudinally to create two pens. The feeding area is 6.9 m wide and therefore covers about 25% of the total barn floor area. The floor of the feeding area is rough concrete with a 1.5% slope toward the exterior of the building and at a height of 0.4 m above the grade of the resting floor. There is no transverse slope.

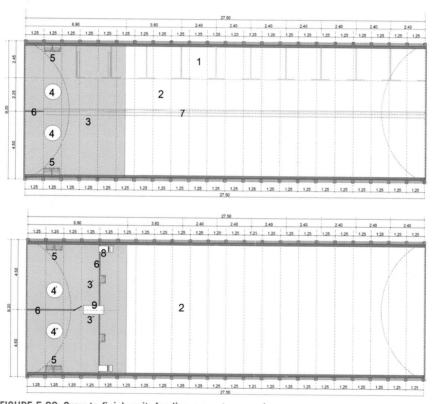

FIGURE 5.20 Grow to finish unit, feeding area at one end.

1. Area for farrowing boxes if used as a farrowing unit; 2. deep bedded resting area; 3. concrete feeding area; 3' & 3". feeding areas for heavy and lightweight pigs, respectively; 4. feeders; 4' & 4". feeders for heavy and lightweight swine; 5. water fountains; 6. optional dividing wall; 7. optional radiant tube; 8. one-way exit gates; 9. electronic sort gate system.

Credit: Hutu, I., Onan, W. G., 2007, Tehnologii Alternative Pentru Creşterea Porcilor, Ed. Mirton, Timişoara -
Swine facilities from River Falls, Wisconsin, USA.

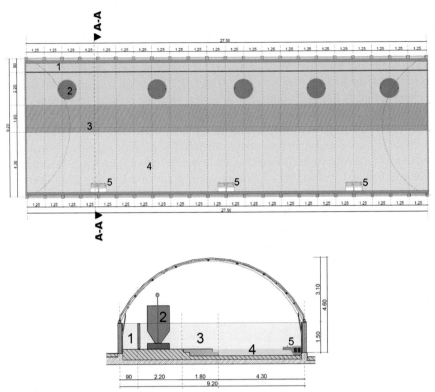

FIGURE 5.21 Floor plan (upper) and cross section (lower) of a grow/finish unit with feeding area on one side.

1. Service alley; 2. feeder; 3. concrete feeding area; 4. deep bedded resting area; 5. heated frost proof water fountain.

Draw after Experimental Units, 2010, Timisoara, RO.

Variant 1 (left side of Fig. 5.20) has two round self-feeders evenly spaced across the floor and two water fountains located at each lateral wall.

Variant 2 of Fig. 5.20 is a design that may be used when an electronic sorting gate is installed. In this case, the resting area is not divided, but the feeding area is divided into an area for lighter weight animals and one for heavier weight animals.

The feeding area is separated from the resting area by pen panels. Access to the feeding areas is provided by the electronic sorting gate, which directs individual animals to the appropriate feeding area based on the animal's weight. Animals exit their feeding areas via one-way gates, leading to the resting area. Water fountains may be located in the feeding area along the pen panel separating it from the resting area. The resting area encompasses the remaining 75% of the barn's total floor area. It may have a concrete floor or a packed clay floor.

FIGURE 5.22 Barns: feeding area at one end (upper) and along one side (lower).

Photo: Ioan Hutu, 2004, River Falls, Wisconsin, USA (upper) and Payne, H., 2004. Deep-litter housing systems an Australian perspective, Hoop Barns & Bedded Systems for Livestock Production Conference, 14 September 2004, Ames, Iowa, Australia (lower). With permission from Honeyman, ISU Hoop Conference.

Hoop structure with feeding area located along one sidewall. Gestation units with feeding stalls along one side can easily be converted to this design by removal of the feeding stalls and installation of appropriate feeders for finishing swine. Fig. 5.21 and the lower portion of Fig. 5.22 illustrate such a design. As usual, the floor area is divided into resting and feeding areas.

The feeding area width is typically 2.9–4.9 m wide, which in a 9- to-20-m-wide barn uses 24%–53% of the available floor area. Feed may be provided using self-feeders located near the south end of the structure or by using feed metering devices spaced along the sidewall with automatic delivery systems (Fig. 5.22, lower). If such automatic systems are used, the feeding area need not be as wide. Again, flooring in the feeding area should be rough concrete at a 0.4 m elevation with a transverse slope of 1%–5%.

FIGURE 5.23 Swine finishing barn.

Upper: view of entire pen. Lower: view of feeding and resting area; note corn stalk bedding bale in foreground and round self-feeder and water fountain in background.

Photo: Ioan Hutu, 2004 (upper) and Gary Onan, 2007 (lower), River Falls, Wisconsin, USA.

Water can be provided either by nipple waterers or by heated frost-proof water fountains. Generally, the entire concrete area along the sidewall of a former gestation unit is not required for adequate feeding space for finishing swine and so a portion of it may be bedded and used as additional resting area. The resting area will cover 76%−47% of the available floor area depending on the width of the feeding area and will be deep bedded.

5.5.2 Group management and time requirements

Depending on the farm's biosecurity protocols, several swine group handling approaches are possible. In the most common scenario, pigs weighing around 15 kg are acquired from a farrowing operation and moved as one group into a grow/finish structure.

The Vastgotmodellen management system has sows farrowing in temperature-controlled barns and piglets remaining in the farrowing unit until 14 weeks of age and about 25 kg. At that age, they are moved to grow/finish hoops, which may be located on the same farm site or at a separate location. These piglets will be maintained as a group and marketed out of the hoop structure at a weight of about 120 kg

FIGURE 5.24 Functional areas in a finishing hoop.

Upper: Waterer in corner of pen (upper left) resting area with bale partly spread and dunging area at the side of the barn (upper right). Lower: Feeding area with round self-feeder and water fountain in barn corner (lower left); resting area with bale partly spread and dunging area in the background at north end of barn (lower right).

Photo: Gary Onan, 2005 (lower), River Falls, Wisconsin, USA and Ioan Hutu, 2016 (upper), Experimental Unit, Timisoara, Romania.

(Honeyman and Kent 1997). Another management system has the piglets remaining in the farrowing unit during the entire grow/finish phase and marketed from that structure at 6—7 months and about 110—115 kg final weight (Harris 2000). In this system, the first 2—3 weeks postweaning is the most critical time, especially in cold weather (Larson et al. 2003). For structures as illustrated in Figs. 5.14 and 5.16 with 20 sows weaning an average of 10 piglets each, 200 finishing hogs will be raised in the hoop and have an available floor space of 1.25 m² each.

It is important that the pigs be sorted by weight and age as they are allocated to different hoops or pens within hoops. Typically, if the pens are populated with homogenous groups and plenty of bedding is supplied, there will be little aggression even with large groups (Arey and Franklin, 1995; Turner et al. 2001). During the grow/finish period, no resorting of groups is typically done. If variant 2 from Fig. 5.20 is used, the electronic sorting apparatus will allow for separate rations for smaller versus larger pigs and thus allow for smaller pigs to catch up with their growth to some extent.

Time required to manage grow/finish units is typically 20–30 h for each cycle of a 200 hog barn (Fulhage, 2003). This includes time spent cleaning and bedding the unit, but not the time required for baling and storing bedding or spreading manure onto fields.

Richard et al. (1998) suggest 20–40 min of labor per hog, whereas Duffy and Honeyman (1999) indicate that with proper labor saving equipment, as little as 15 min (0.25 h) is required per head.[15] Daily observation of animals, cleaning, feeding, and sorting take up about 60% of the total time required to raise the hogs. Other chore time requirements will depend significantly on the level of mechanization available.

5.5.3 Management of barn environment

Hoop structures designed for grow/finish stages of swine production are the simplest with merely two functional areas established by sectioning the building either transversely or longitudinally. Typically, two feeders are provided in the feeding area. They may be either round or rectangular self-feeders (Figs. 5.22 and 5.23) with a minimum of 12 feeding spaces each.

Ventilation is provided by natural air flow. Weaned piglets require 20 m^3 of air flow per hour per animal. This increases to 60 m^3 per hour for growers and 90 m^3 per hour for finishing hogs.

Temperature in these structures is dependent on ambient temperature. For 15 kg piglets, the ideal temperature range is 20–22°C, whereas for 25 kg piglets, the ideal temperature is 18–20°C (Malachy, 2006). If temperatures are outside these ranges, there will be effects on feed intake and potentially, on respiratory health. At temperatures below the lower critical temperature,[16] feed intake will increase by 5 g per kg of hog weight for each degree Celsius. Sallvik and Weifeldt (1993) indicate that for every 1° rise above 15°C, feed intake drops by 35 g per day per degree between 15 and 20°C and by 75 g per day per degree between 20 and 25°C. Temperature effects on productivity vary with body weight, being more pronounced for heavier weight animals (Malachy 2006). For finishing swine, average daily gain drops by 10–12 g for each degree Celsius below the lower critical temperature (www.thepigsite.com).

Lighting should include at least 8 h of light per day with a minimum intensity of 50–100 lumen/m^2 for growing pigs. Longer light intervals will increase growth rate and feed intake (Bruininx et al. 2002). For finishing hogs, again a minimum of 8 h at 50 lumens/m^2 intensity is recommended.

---

[15] Fifteen minutes/animal: 1.8 min (108 s) for bedding; 3.0 min (180 s) for daily observation; 3.6 min (216 s) for feeding; 0.6 min (36 s) for sorting; 3.6 min (216 s) for veterinary care; and 0.6 min (36 s) for miscellaneous duties.

[16] When swine are in large groups on deep bedding, the lower critical temperature is about 10°C for 20 kg pigs, 8°C for 60 kg pigs, and 2°C for large hogs of above 100 kg (Draghici 1991).

Bedding usage will depend on the type of material used and on the season (Table 3.3). During winter, about 150 kg per head is required, and during summer, about 100 kg per head. Typically for a group of 200 grow/finish swine, 5−8600 kg bales will be used for the initial bedded layer with about 2 additional bales added per week. These subsequent bales need not be spread by the manager as the hogs will very effectively spread them themselves.

5.5.4 Grouping and space requirements

The swine may be placed into two separate longitudinal pens, 70−100 head each, with heavier ones on one side and lighter animals on the other. Conversely, they may be left in one large group of 150−200 with no subdivision of the structure. In this case, however, use of an electronic sorting system (variant 2 in Fig. 5.20) may give better performance as it will allow for different rations for the lighter and heavier pigs.

The sorting system will allow individuals access to one of the two feeding areas with differing rations. The sort is based on animal weight or weight in relation to the group average. In general, however, if animals are of uniform age and weight, large groups work well from behavioral and performance aspects.

Space allotment per animal is of course a function of age and weight. Some typical recommended values are as follows:

- 0.55 m^2 per piglet up to 10 weeks of age or 35 kg (Kruger et al. 2006);
- 1.00 m^2 per growing pig up to 20 weeks of age (Kruger et al. 2006; Brumm et al. 2004);
- 1.10 m^2 per finishing hog up to 24 weeks age or 110 kg (Kruger et al. 2006; Brumm et al. 2004) or 1.5 m^2 per head for organic farms.

To mitigate aggression and decrease biting injuries in groups of young swine, one adult boar may be introduced into the group (Grandin and Bruning, 1999). One feeding space should be allowed for every 79 head (Morrison and Lee 2003). Pigs that are ill or injured should be removed from the group. Farms should have a special needs area to house such animals. Furthermore, these animals should not be reintroduced into the group as they are likely to suffer from aggressive action by established pigs in the group.

5.5.5 Health management

Health practices vary depending on the biosecurity protocols in place at a particular farm site. In general, however, some common practices would include deworming (with ivermectin) when piglets are 15−16 kg and vaccination for atrophic rhinitis. The initial dose of rhinitis vaccine is generally administered at 3 weeks of age and the booster dose 1 month later. Young boars and gilts should be vaccinated at 6 month intervals (Moga 2001). Oftentimes when the swine reach 50−60 kg, another dose of anthelminthic may need to be administered. A product from the benzimidazole family may be advised at this time (Honeyman; Harmon 2003). In situations where aggressive behavior is a problem upon moving the group into the hoop, some drug therapies may be advantageous as well. Odor masking agents (OMAs) may be used to block the animals' ability to recognize other individuals via this important sense (Friend et al. 1983; Krebs; McGlone 2004).

In severe instances, tranquilizers may be utilized to reduce aggression as well (Gonyou et al. 1988).

5.5.6 Nutrition management

Feed is supplied ad libitum with self-feeders and is generally in dry meal form ground to a particle size of 650–750 μm (Honeyman; Harmon 2003). Daily feed intake varies with the growth stage of the swine and the season of the year. Average intake over the entire growth period will be about 2.3 kg per day during warm seasons and 2.5 kg per day during cold seasons (Larson et al., 1999a, 1999b). Modern practice in grow/finish nutrition utilizes a concept known as phase feeding where the pigs are offered a series of different diets as they grow and develop. The diet for each phase is specifically formulated to meet the particular nutritional requirements of the animal at that time based on its age, weight, expected performance ability, sex, and season of the year. The nutrient requirements for piglets from weaning to 25 kg are as follows:

- 5–7 kg piglets raised in 25°C housing will have a daily requirement of 904 kcal. ME, 61 g total protein (of which 4.5 g is lysine), 2.26 g Ca, 1.20 g available P, and 1.60 mg Cu; when the ration density is 3300 kcal/kg ME (3438 kcal/kg DE), the animals will consume about 0.28 kg of the ration per day.
- 7–11 kg pigs maintained at 20°C have a daily requirement of 1592 kcal. ME, 96 g total protein (of which 7.2 g is lysine), 3.75 g Ca, 1.87 g available P, and 2.81 mg Cu; when the ration density is 3300 kcal/kg ME, these animals will consume about 0.49 kg/day.
- 11–25 kg pigs maintained at 15°C will require 3033 kcal. ME, 171 g total protein (12.6 g lysine), 6.34 g Ca, 3 g available P, and 4.53 mg Cu; those would require intake of 0.95 kg of a 3300 kcal./kg ME ration per day.

For Romanian farms, Luca (2000) suggests that for piglets up to 25–30 kg, a feed ration with 3100–3200 kcal/kg ME, 18% DP, 0.75% lysine, 0.56% methionine + cystine, 0.7% Ca, 0.5% digestible P, and a maximum of 4.0% fiber be used.

Daily consumption of this ration will vary from 0.7 to 1.5 kg per day (4%–5% of body weight) over this time period of the pigs' growth. Such a ration can be achieved with the following ingredient quantities:

- Energy feeds: 40%–50% corn; 10%–20% barley, oats, or wheat; 1.5%–4% fat; and 5%–10% wheat bran.
- Protein feeds: 15%–20% sunflower or soybean meal; 8%–10% animal protein (fish meal or dried blood plasma); and 3%–4% dried whey.

- Minerals: 1%−1.5% limestone (feed lime); 0.5%−1.0% phosphates; 0.5% salt; and 1.0% vitamin/trace mineral premix with 250 ppm Cu.
- Feed additives/growth promotants: prebiotics and/or probiotics and/or acidifiers for maintenance of gut health.

For pigs weighing between 20 and 120 kg, the total expected feed intake to gain 95 kg (from 25 to 120 kg) would be 315−335 kg (Table 5.8). This equates to 3.3−3.5 kg/day average over the growth period (Brumm et al. 1997). Daily nutrient requirements (NRC, 2012) for pigs with current hybrid genetics is as follows (this assumes a ration with 3300 kcal/kg ME):

- 25−50 kg pigs: 4959 kcal ME, 236 g total protein of which 16.9 g is lysine, 9.87 g Ca, and 4.6 g available P. Daily feed intake will be about 1.58 kg.
- 50−75 kg pigs: 6989 kcal. ME, 291 g total protein of which 20.6 g is lysine, 12.43 g Ca, and 5.8 g available P. Daily feed intake will be about 2.23 kg.

Table 5.8 Example rations and expected performances for grow/finish swine at various weights.

Ingredient	Body weight (kg)			
	25−50	50−75	75−100	100−135
Corn, %	67.30	72.5	72.70	73.75
Barley, %	−	−	5	8
Soybean meal, %	30.00	25.00	20.00	16.00
Dicalcium phosphate, %	1.20	1.00	0.75	0.70
Feed lime, %	0.70	0.70	0.75	0.80
Salt, %	0.40	0.40	0.40	0.35
Vitamin/trace mineral premix, %	0.40	0.40	0.40	0.40
Ration analysis				
Metabolizable energy, kcal/kg	3274	3286	3281	3277
Total protein, %	19.87	17.92	16.11	14.63
Lysine, %	0.93	0.80	0.69	0.59
Tryptophan, %	0.21	0.18	0.16	0.14
Threonine, %	0.62	0.55	0.49	0.43
Methionine + cysteine, %	0.56	0.51	0.47	0.44
Total calcium, %	0.74	0.68	0.62	0.61
STTD phosphorus, %	0.37	0.33	0.28	0.27
Fermentable fiber, %	12.15	11.26	10.57	9.97
Expected consumption and performance				
Expected average ME intake, kcal/day	4961	6992	8267	9198
Expected daily intake, kg/day	1.52	2.13	2.52	2.81
Expected average daily gain, g/day	759	900	917	867

After NRC 2012 requirements, calculated for gilts and barrows in thermoneutral environment and feeds with the indicated energy and fermentable fiber content.

- 75–100 kg hogs: 8625 kcal. ME, 304 g total protein of which 21.1 g is lysine, 13.14 g Ca, and 6.1 g available P. Daily feed intake will be about 2.64 kg.
- 100–135 kg hogs: 9196 kcal. ME, 291 g total protein of which 19.7 g is lysine, 12.8 g Ca, and 5.95 g available P. Daily feed intake will be about 2.93 kg.

Romanian farms growing pigs for multinational swine companies are typically feeding rations with up to seven phases (seven different nutrient periods as outlined in previous paragraphs). These rations would consist of energy feeds such as corn or barley and protein supplements such as soybean or rape meal with some added crystalline amino acids.

On Romanian farms, grow/finish swine are typically fed using two ration phases,[17,18] which may be formulated from the following:

- Energy feeds: 60%–75% corn, 10%–30% barley (especially during final finishing phase), and 5% wheat bran.
- Protein feeds: 8%–13% soybean/sunflower meal or 10% field peas.
- Animal protein feeds: 3%–6% meat and bone meal or blood meal.
- Mineral feeds: 1% feed lime, 0.5% phosphates, 0.5% salt, 0.05% copper sulfate, and 1% vitamin/trace mineral premix.

5.5.7 Expected performance

Grow/finish performance will depend on the breed or hybrid used. Hybrids will have the advantage of heterosis (hybrid vigor), which will increase their growth rate and feed efficiency. Table 5.9 tabulates production characteristics of a Hampshire, Yorkshire, Duroc cross, and PIC hybrid line when used in a hoop rearing system.

As indicated by the results of Canadian research (Connor, 1993, 1994; Connor et al., 1994, 1997; Matte, 1993) and trials from Iowa (Larson et al., 1999a, 1999b; Larson and Honeyman, 1999, Table 5.10; Honeyman and Harmon, 2003), swine grown in cold hoop barns performed favorably. There are seasonal differences within the hoop system. For example, when compared with the average annual feed consumption, that for winter is 5%–9% higher and that for summer is 5%–9% lower. Average daily gain may be 10%–15% higher during summer compared with the annual average. Swine finished during the summer may reach market 1 month sooner than those finished during winter. Summer/winter differences vary considerably depending on climate and weather. Bedding use is 33% higher in winter, and total manure production is 2.7 times higher compared with the annual average. Swine finished in cold hoop structures have advantages and disadvantages

---

[17] The earlier phase (from 25–30 kg piglets up to 50–60 kg pigs) requires a ration with 3050–3150 kcal/kg ME, 16% DP, 0.7% lysine, 0.5% methionine + cystine, 0.65% Ca, 0.4% P, and 0.5% salt with a maximum of 5% fiber.

[18] The ration density for the latter phase (from 50–60 kg piglets up to 110–120 kg pigs) should be 3150 kcal/kg ME, 13%–14% DP, 0.55% lysine, 0.43% methionine + cystine, 0.4% Ca, 0.5% P, and 0.5% salt with a maximum of 6% fiber.

Table 5.9 Values for production traits of swine finished in hoop barns.

Specification	Unit	Finishing over	
		Winter	Summer
Hybrid used		HxYxD	PIC
Start date		16th November	15th April
Market date		February–March	August–September
Average days on feed	days	108	122
Marketing span[a]	days	28	31
Available floor area (9.20 × 18.50 m)	m^2	170.2	170.2
Number placed on feed	animals	151	150
Space per animal	m^2	1.13	1.13
Initial weight	kg	25	23.1
Number marketed	animals	147	146
Mortality loss	%	2.6%	2.7%
Weight at market	kg	111.6	120.2
Hot carcass weight	kg	83.9	90.7
Dressing percent	%	75.2%	75.5%
Percent fat free lean	%	46.7	47.6
Back fat thickness	mm	25.7	25.1
Total gain	kg	86.6	97.1
Average daily gain	g/day	802	795
Feed:gain	kg feed/kg gain	3.53	3.27
Feed consumption/pig marketed	kg/pig	306	317
Feed intake/pig/day	kg/pig/day	2.83	2.60
Bedding use	ton	17.96	8.44
Bedding/pig marketed	kg	118.8	56.2
Manure production	ton	100	36
Labor/pig marketed[b]	minutes/seconds	36'44"	25'28"

[a] Time in days between sales of first and last animals from each lot.
[b] Labor covers daily inspection, addition of deep litter bedding, sorting and loading of market swine, removal and disposal of manure; it does not cover feed preparation and distribution, or transport of pigs to market.
After Honeyman, M. S., 1996. Sustainability issues of US swine production, J. Anim. Sci. 74, 1410. 1217.

Table 5.10 Comparative performance of swine grown and finished in hoop barns (HB) versus conventional (CV).

Growth season	Summer		Winter	
Type barn: hoop/conventional	**HB**	**CV**	**HB**	**CV**
Growth performance				
Number of swine	451	132	451	132
Initial weight, kg	16	17	14	15
Market weight, kg	118	118	119	116
Total gain, kg	102.6	100.8	104.2	101.2
Days on feed	117	122	148	136
Bedding use, kg	88.5	-	99.8	-
Feed intake, kg/day	2.4	2.4	2.3	2.2
ADG, g/day	866	826	712	744
Feed:gain, kg feed/kg gain	2.8	3.0	3.3	3.0
Mortality, %	2.0	4.5	5.5	2.3
Involuntary culls, %	0.0	0.0	3.5	2.3
Performance at ultrasound test				
Test age, days	113	118	132	125
Test weight, kg	112.1	115.4	105.9	107.8
Fat thickness, mm	24.6	21.6	19.6	19.6
LMA, cm^2	38.1	41.4	38.7	42.3
Lean yield, kg/swine	41.6	44.6	41.5	43.2
Fat free lean, %	50.2	52.2	53.1	54.3

Adaptation and compilation after Larson, E., et al., 1999a, Iowa State University, USA trials.

in performance compared with those raised in conventional units (Lay et al. 2000; Johnston et al. 2006). Some examples are as follows:

- Daily feed consumption is 6% lower in summer and 10% higher in winter than in conventional units.[19]
- Average daily gain is 5% higher in summer and 4% lower in winter compared to conventional units.
- Fat depth is about 0.4 mm thicker (2.4%) for winter fed hogs (Gentry et al., 2002; Gentry and Johnson, 2004).

---

[19] Magolski and Onan (2007) using data from a 3-year comparison between hoop and conventionally raised swine conducted in Wisconsin where the climate is similar to Romania's concluded that there is a strong trend for higher feed:gain ratios for hoop hogs versus confinement hogs in the winter (3.69 vs. 3.4 kg feed/kg gain with P = .057). The same authors indicate that hoop-finished swine in winter had thicker fat depth (21.2 vs. 16.3 mm, P < .01) and lower percent fat free lean (52.5% vs. 55.2%, P < .01) than confinement-raised swine.

When results are combined over all seasons, the differences between cold and conventional units are as follows (Larson et al., 1999a):

- Daily feed consumption—hoop barn value is higher by 2% (2.39 vs. 2.34 kg).
- Average daily gain: hoop barn swine gain is about 1% faster (794 vs. 785 g/day);
- Feed/gain: hoop barn swine required are about 3% more feed per unit gain (3.05. vs. 2.97);
- Fat depth: hoop barn raised swine have about 7% thicker fat layer (22.1 vs. 20.6 mm).
- Loin muscle area: hoop barn finished swine have an 8% decrease (38.5 vs. 41.9 cm^2).
- Percent fat free lean: hoop barn swine are lower by about 3% (51.7 vs. 53.3%).

In summary, it seems that swine raised in cold hoop structures gain a bit faster but are a bit fatter at market weight than those raised in conventional confinement systems (Brumm et al., 1997; Honeyman and Harmon, 2003; Larson et al., 1999a).

In a trial conducted between April and August of 2016 in Timisoara, Romania at Banat University of Agricultural Science and Veterinary Medicine "King Michael I of Romania" (Hutu and Onan 2016; Onan et al. 2016) commercial swine obtained from Smithfield Romania were grown in a hoop style finishing unit (Fig. 5.25). The animals consisted of 20 barrows with a mean starting weight of 28 kg and 21 gilts with a mean starting weight of 27 kg. Table 5.11 illustrates key outcomes from the trial.

Mean ambient temperature for the duration of this trial was 19.95°C (range: 13.98−26.01°C). Average daily gain for barrows and gilts combined was 913 g/day (range: 670−1023 g/day).

Table 5.11 Comparative performance of barrows versus gilts in a summer hoop feeding trial in Romania.

Item	Barrows	Gilts	P Value
Average daily gain (g)	936	888	0.063
Feed conversion (feed:gain)	2.71	2.66	0.997
Slaughter weight (kg)	114	109	0.049
Carcass weight (kg)	87.9	83.7	0.112
P2 fat depth (mm)	14.5	12.3	0.011
Eye muscle depth (mm)	58.6	53.7	0.019
Percent lean	58.8	59.9	0.126

After Hutu, I., Onan, G., 2016. Hoop structure for wean to finish pigs: Gender Influences on Carcass in a Summer Trial. College of Agricultural Science and Veterinary Medicine, Iasi, RO: Veterinary Conference, vol. 59 (fourth ed.). pp. 381−387.

FIGURE 5.25 Swine Experimental Unit.

Swine hoop at Horia Cernescu Research Unit[20] during summer season.

Photo: Hutu, 2016, Experimental Units, Timisoara, Romania.

This trial confirms that hoop structures work well for grow-finish swine production under Romanian summertime conditions as these growth and carcass measures are comparable with those obtained in conventional units.

References

Algers, B., 1991. Group housing of farrowing sows, health aspects of a new system. In: Proceedings Seventh International Congress on Animal Hygiene, Leipzig, Germany, pp. 851–857.

Arey, D.S., 1992. Straw and food as reinforcers for prepartal sows. Appl. Anim. Behav. Sci. 33, 217.

Arey, D.S., Franklin, M.F., 1995. Effects of straw and unfamiliarity on fighting between newly mixed growing pigs. Appl. Anim. Behav. Sci. 45, 23–30.

Baker, R.B., 2004. The application of biosecurity protocols in large production systems: experiences from the field. In: Leman, A.D. (Ed.), Swine Conference, pp. 90–91.

Baxter, M.R., 1991. The freedom farrowing system. Farm Build. Progr. 104, 9.

Beltranena, E., Patterson, J., Willis, J., Foxcroft, G., 2005. The boar exposure area of the GDU (Gilt Development Unit). Western Hog J. 27 (1), 29–31.

Beltranena, E., Patterson, J., Foxcroft, G., 2005a. Optimizing boar power use in the gilt development unit. Western Hog J. 27 (1), 33–36.

Braun, S., Algers, B., 1993. Schweden-stall für grosse altgebäude. DLG-Mitteilungen/Agrarinform 4, 60–61.

[20] The activities and materials were sponsored under the framework of Smithfield Romania University Program, and the research was carried out within the Swine Experimental Unit, a part of Horia Cernescu Research Unit at the Banat University of Agricultural Science and Veterinary Medicine "King Michael I," infrastructure established by POSCCE project—Development of research, education and services infrastructure in the fields of veterinary medicine and innovative technologies for West Region—contract no. 18/March 1, 2009; 2669 SMIS code.

Bruininx, E.M., Heetkamp, A.M., Bogaart, D., Peet-Schwering, C.M.C., Beynen, A.C., Evert, H., Hartog, L.A., Schrama, J.W., 2002. A prolonged photoperiod improves feed intake and energy metabolism of weaned pigs. J. Anim. Sci. 80, 1736, 174.

Brumm, M.C., Harmon, J.D., Honeyman, M.S., Kliebenstein, J.B., 1997. Hoop Structures for Grow-Finish Swine. Iowa State Univ., Ames, IA. MidWest Plan Service, AED-41.

Brumm, M.C., Harmon, J.D., Honeyman, M.S., Kliebenstein, J.B., Lonergan, S.M., Morrison, R., Richard, T., 2004. Hoop Barns for Grow-Finish Swine, Rev. Ed. MidWest Plan Service, Iowa State Univ., Ames, IA, p. 24. AED41.

Brüssow, K.P., Egerszegi, I., Ratky, J., Torner, H., Toth, P., Schneider, F., 2005. Reproduction in the Hungarian mangalica pig: review article. Pig News Inf. 26 (1), 23N–28N.

Christensen, G., Vraa-Andersen, L., Mousing, J., 1995. Causes of mortality among sows in Danish pig herds. Vet. Rec. 137, 395–399.

Connor, M.L., 1993. BioTech shelters, alternative housing for feeder pigs. Manitoba Swine Seminar. Proc. 7, 81.

Connor, M.L., 1994. Update on alternative housing for pigs. Manitoba Swine Seminar Proc. 8, 93–96.

Connor, M.L., Onischuk, L., Zhang, Q., Parker, R.J., Elliot, J.I., 1994. Alternative housing with canadian biotech shelters and a review of some European concepts. Canadian Society of Agr. Engineering 1994 proceedings.

Connor, M.L., Fulawaka, D.L., Onischuk, L., 1997. Alternative low-cost group housing for pregnant sows. In: Livestock Environment. Proceedings of the Fifth International Symposium ASAE, St, Joseph, MI, vol. 1.

Crăiniceanu, E., Matiuti, M., 2006. In: Nutritia Animalelor. Brumar. Timişoara.

Culbertson, M., 2014. Feeding sows for maximum lifetime production. In: Lehman, A.D. (Ed.), Swine Conference, St. Paul, MN USA.

Dinu, I., Hălmăgean, P., Tărăboantă, G., Farkaş, N., Simionescu, D., Popovici, F., 1990. Tehnologia creşterii suinelor. In: Didactică şi Pedagogică, Bucureşti.

Docking, C.M., Kay, R.M., Day, J.E.L., Chamberlain, H.L., 2001. The effect of stocking density, group size and boar presence on the behaviour, aggression and skin damage of sows in a specialized mixing pen at weaning. In: Proceedings of the British Society of Animal Science, p. 46.

Drăghici, C., 1991. In: Microclimatul adăposturilor de animale şi mijloacele de dirijare. Ceres, Bucuresti.

Duffy, M.D., Honeyman, M., 1999. Labor requirements for market swine produced in hoop structures Iowa state University management/economics. In: 1999 ISU Swine Research Report. Iowa State University, Ames. ASL-R1685.

Durrel, J., Sneddon, I.A., Beattie, V.E., 1997. Effects of enrichment and floor type on behaviour of loose-housed dry sows. Anim. Welf. 6, 297–308.

D'Allaire, S., Morris, R.S., Martin, F.B., Robinson, R.A., Leman, A.D., 1989. Management and environmental factors associated with annual sow culling rate: a path analysis. Prev. Vet. Med. 7, 255–265.

Ehlorsson, C.J., Olsson, O., Lundeheim, N., 2002. Inventering av klövhälsan hos suggor i olika inhysningsmiljöer [trad. Investigations of housing and environmental factors affecting the claw health in group-housed dry sows]. Svensk Veterinärtidning 54, 297–304.

Ewan, R.C., Crenshaw, T.D., Cromwell, G.L., Easter, R.A., Nelssen, J.L., Miller, E.R., Pettigrew, J.E., Veum, T.L., 1996. Effect of addition of fibre to gestation diets on reproductive performance of sows. J. Anim. Sci. 74 (Suppl. 1), 190.

Fisher, D.M., 1990. The application of electronic identification to groups of farrowing and lactating sows in straw bedded housing. In: Electronic Identification in Pig Production. RASE, Stoneleigh, p. 101.

Friend, T.H., Knabe, D.A., Tanksley, T.D., 1983. Behavior and performance of pigs grouped by three different methods at weaning. J. Anim. Sci. 57, 1406–1411.

Fulhage, C., 2003. Manure Management in Hoop Structures Nutrients and Bacterial Waste. MU Extension, University of Missouri-Columbia, muextension.missouri.edu/xplor. Agricultural Engineering Extension, EQ 352.

Gallois, M., le Cozler, Y., Prunier, A., 2005. Influence of tooth resection on piglets welfare and performance. Prev. Vet. Med. 69, 13–23.

Gentry, J.G., McGlone, J.J., Blanton Jr., J.R., Miller, M.F., 2002. Alternative housing systems for pigs: influences on growth, composition, and pork quality. J. Anim. Sci. 80, 1781–1790.

Gentry, J.G., Johnson, A.K., 2004. Behavior of finisher pigs housed in hoop and conventional buildings: how does this affect pork quality and eating experience? In: Alternative Swine Housing Systems: An International Scientific Symposium, September 15, 2004, Ensminger Room, Kildee Hall, Iowa State University Ames, Iowa.

Gonyou, H., 2003. Group housing: alternatives systems, alternative management. Adv. Pork Prod. 14, 101–107.

Gonyou, H.W., Rohde-Parfet, K.A., Anderson, D.B., Olson, R.D., 1988. Effects of amperozide and azaperone on aggression and productivity of growing-finishing pigs. J. Anim. Sci. 66, 2856–2864.

Gourdine, J.L., Renaudeau, D., Noblet, J., Bidanel, J.P., 2004. Effects of season and parity on performance of lactating sows in a tropical climate. Anim. Sci. 79, 273–282.

Grandin, T., Bruning, J., 1999. Boar presence reduces fighting in mixed slaughter-weight pigs. Appl. Anim. Behav. Sci. 33, 273–276.

Halverson, M., 1991a. Farm Animal Welfare: Crisis or Opportunity for Agriculture? Dept, of Agriculture and Applied Economics, University of Minnesota, St. Paul, p. 68. Staff paper 91-1.

Halverson, M., 1991b. Using Nature as Both Mentor and Model: Animal Welfare Research and Development in Sustainable Swine Production. Dept, of Agriculture and Applied Economics, University of Minnesota, St. Paul, p. 44. Staff paper P 91-26.

Halverson, M., 1994. Whole-hog housing Swedish system lowers stress, disease. The New Farm 16 (2), 51–54, 62.

Halverson, M.K., Honeyman, M.S., 1997. Swedish Deep-Bedded Group Nursing System for Feeder Pig Production. Swine Systems Options for Iowa No. SA-12. Iowa State University Extension, Ames.

Halverson, M., Honeyman, M.S., Adams, M., 1997. Swedish Deep-Bedded Group Nursing Systems for Feeder Pig Production. Iowa State University Extension, Ames, IA. AS-12.

Harmon, J.D., Honeyman, M.S., Kliebenstein, J.B., Richard, T., Zulovich, J.M., 2004. Hoop Barns for Gestating Swine, Rev. Ed. MidWest Plan Service. Iowa State Univ., Ames, IA, p. 20. AED44.

Harris, D.L., 2000. Multi–site Pig Production. Iowa State Univ. Press, Ames, Iowa.

Hilbrands, A., Johnston, L., Morrison, R., Jacobson, L., 2004. Conversion of a partially slatted gestation barn to a deep bedded farrowing system. In: Hoop Barns & Bedded Systems for Livestock Production Conference, 14 September 2004, Ames, Iowa. West Central Research and Outreach Center, Morris, MN.

Holden, P., Ewan, R., Jurgens, M., Stahly, T., Zimmerman, D., 1996. ISU Life Cycle Swine Nutrition PM-489. Iowa State Univ, Ames, IA, p. 17.

Honeyman, M.S., 1995. Västgötmodellen Sweden's sustainable alternative for pig production. Am. J. Altern. Agric. 10 (3), 129—132.

Honeyman, M.S., 1996. Sustainability issues of US swine production. J. Anim. Sci. 74, 1410, 1217.

Honeyman, M.S., 1997. Hooped Structures with Deep Bedding for Grow-Finish Pigs. Iowa State University Extension, Ames, IA, pp. 107—110. ASL-R1392. 1996 Swine Research Reports.

Honeyman, M.S., Harmon, J.D., 2003. Performance of finishing pigs in hoop structures and confinement during winter and summer. J. Anim. Sci. 81, 1663—1670.

Honeyman, M.S., Johnson, C., 2002. Sow and Litter Performance of Individual Crate and Group Hoop Barn Gestation Housing Systems: A Progress Report. Iowa State University, Armstrong Research and Demonstration Farm ISRF 02-12.

Honeyman, M.S., Kent, D., 1996. Demonstrating a Swedish Feeder Pig Production System in Iowa. Iowa State Univ. Ext., Ames. ASL- R1390. Swine Research Report.

Honeyman, M.S., Kent, D., 1997. First Year Results of a Swedish Deep-Bedded Feeder Pig Production System in Iowa. ISU Extension, Ames. Swine Res. Rept ASL-R1497.

Honeyman, M.S., Kent, D., 2000. Reproductive Performance of Young Sows from Various Gestation Housing Systems, 1999 Swine Research Reports. Iowa State University, Ames, IA. AS-642.

Honeyman, M.S., Kent, D., 2001. Performance of a Swedish deep-bedded feeder pig production system in Iowa. Am. J. Altern. Agric. 16 (2), 50—56.

Honeyman, M.S., Harmon, J., Lay, D., Richard, T., 1997. Gestating sows in deep-bedded hoop structures ASL-R1496 Management/Economics Iowa State. In: 1997 ISU Swine Research Report, AS-638. Dept, of Animal Science, Iowa State University, Ames.

Honeyman, M.S., Harmon, J.D., Kent, D., Hummel, D., Christian, L., 2002. The Effects of Gestation Housing on the Reproductive Performance of Gestating Sows: A Progress Report ISRF01-12. Iowa State University, Armstrong Research and Demonstration Farm.

Houwers, H.W.J., Bure, R.G., Koomans, P., 1992. Behaviour of sows in a free-access farrowing section. Farm Build. Progr. 109, 9.

Houwers, H.W.J., Bure, R.G., Walvoort, J., 1993. Production aspects of integrated housing of sows with confined litters. Anim. Prod. 56, 477.

Hutu, I., 2004a. Herd reproduction disease. Concept and importance in Intensive reproduction management in sows. Rev. Rom. Med. Vet. 14 (2), 31—50.

Hutu, I., 2004b. Reproduction herd disease and its risk factors in intensive reproduction management of sow. Buletin USAMV-CN 60, 201—206.

Hutu, I., 2005. In: Reproducerea intensivă la scroafe. Mirton, Timişoara.

Hutu, I., Ghesa, I., 2002. Researches on some indicators and factors of technological risk associated with first mating age in Large White gilts, Lucrări Ştiintifice, Zoot. Şi Biotehnologii 35, 269—273.

Hutu, I., Onan, W.G., 2007. In: Tehnologii Alternative Pentru Creşterea Porcilor. Mirton, Timişoara.

Hutu, I., Onan, G., 2016. Hoop structure for wean to finish pigs: gender influences on carcass in a summer trial. In: College of Agricultural Science and Veterinary Medicine, Iasi, RO: Veterinary Conference, fourth ed., vol. 59, pp. 381—387.

Jensen, P., 1986. Observations on the maternal behaviour of free-ranging domestic pigs. Appl. Anim. Behav. Sci. 16, 131.

Johnston, L., Morrison, R., Rudstrom, M., 2006. Welfare, meat quality, growth performance and economics of pigs housed in hoop barns. Western Hog J. 28 (1), 38–44.

Karlen, G.A., Hemsworth, P.H., Gonyou, H.W., Fabrega, E., Strom, D., Smits, R.J., 2005. The Welfare of Gestating Sows in Conventional Stalls and Large Groups on Deep Litter. Report to Australian Pork Limited. Project number 1825.

Kemp, B., Soede, N.M., 1996. Relationship of weaning-to-estrus interval to timing of ovulation and fertilization in sows. J. Anim. Sci. 74, 944–949.

Knox, R.V., Miller, G.M., Willenburg, K.L., Rodriguez-Zas, S.L., 2002. Effect of frequency of boar exposure and adjusted mating times on measures of reproductive performance in weaned sows. J. Anim. Sci. 80, 892–899.

Koketsu, Y., Dial, G.D., 1997. Quantitative relationships between reproductive performance in sows and its risk factors. Pig News Inf. 18, 47N–51N.

Krebs, N., McGlone, J.J., 2004. Maternal pheromone application before and/or after weaning: effects on pig behavior and performance. J. Anim. Sci. 82 (Suppl. 1), 278.

Kruger, I., Taylor, G., Roese, G., Payne, H., 2006. Deep-litter Housing for Pigs, PRIMEFACT 68. NSW Department of Primary Industries, State of New South Wales. http://www.dpi.nsw.gov.au.

Lammers, P., Honeyman, M.S., 2004. Effects of bedded gestation housing on litter size and culling: preliminary results. In: Hoop Barns & Bedded Systems for Livestock Production Conference, 14 September 2004. Iowa, Ames.

Lammers, P.J., Honeyman, M.S., 2005. Farrow-to-Finish in a Hoop Barn: A Demonstration. Animal Industry Report: AS 651, ASL R2029. Available at: https://lib.dr.iastate.edu/ans_air/vol651/iss1/29.

Lammers, P., Honeyman, M.S., Harmon, J., 2004a. Alternative systems for farrowing in cold weather. Agricultural Engineers Digest. AED 47, September, 2004, Available at: http://www.nlfabric.com/assets/documents/alternative-farrowing-systems.pdf.

Lammers, P., Honeyman, M.S., Harmon, J., 2004b. Alternative Winter Farrowing Demonstration Project: Two-Year Summary. Iowa State University, p. 31. Northwest Research Farms and Allee Demonstration Farm ISRF04-29.

Lammers, P., Honeyman, M., Mabry, J., 2006. Sow and Litter Performance for Individual Crate and Group Hoop Barn Gestation Housing Systems: Progress Report III. Iowa State University Animal Industry Report, 2006.

Larson, M.E., Honeyman, M.S., 1999. Performance of Pigs in Hoop Structures and Confinement during Summer with a Wean-To-Finish System ASL-R1681. ISU Ext. Serv., Ames, IA. Swine Research Report.

Larson, M.E., Honeyman, M.S., Penner, A.D., 1999a. Performance of Finishing Pigs in Hoop Structures and Confinement during Summer and Winter. Iowa State Univ., Ames pp. 102–104. Tech. Rep. ASL-R 1682.

Larson, M.E., Honeyman, M.S., Penner, A.D., Harmon, J.D., 1999b. Performance of Finishing Pigs in Hoop Structures and Confinement during Summer and Winter. Iowa State Univ., Ames, pp. 102–104. Tech. Rep. ASL-R 1682.

Larson, M.E., Honeyman, M.S., Harmon, J.D., 2003. Performance and behavior of early–weaned pigs in hoop structures. Appl. Eng. Agric. 19 (5), 591–599.

Lay Jr., D.C., Haussman, M.F., Daniels, M.J., 2000. Hoop housing for feeder pigs offers a welfare-friendly environment compared to a non-bedded confinement system. J. Appl. Anim. Welf. Sci. 3 (1), 33–48.

Leaflet, A.S., 2006. Sow and Litter Performance for Individual Crate and Group Hoop Barn Gestation Housing Systems, p. R2171. Progress Report III Iowa State University Animal Industry Report 2006.

Levis, D.G., 2015. Gilt Management in the BEAR System, Pork Information Gateway, Factsheet. University of Nebraska. http://porkgateway.org/wp-content/uploads/2015/07/gilt-management-in-the-bear-system1.pdf.

Luca, I., 2000. In: Curs de alimentatia animalelor. Marineasa, Timişoara.

Magolski, J., Onan, G.W., 2007. Growth and Carcass Parameters of Hogs Finished in a Deep-Bedded Hoop Structure vs. A Confinement Structure. American Society of Animal Science Midwestern Branch, p. 105 (Abstr.).

Malachy, Y., 2006. Managing growing-finishing pig diets for maximum profitability. Western Hog J. 27 (3), 36–42.

Matte, J.J., 1993. A note on the effect of deep-litter housing on the growth performance of growing finishing pigs. Can. J. Anim. Sci. 73, 642–647.

Matte, J.J., Robert, S., Girard, C.L., Farmer, C., Martineau, G.P., 1994. Effect of bulky diets based on wheat bran or oat hulls on reproductive performance of sows during their first two parities. J. Anim. Sci. 72, 1754–1760.

McGlone, J.J., Morrow, J.C., 1988. Reduction in pig antagonistic behaviour by and osterone. J. Anim. Sci. 66, 880–884.

Moga-Mânzat, R., 2001. Boli infectioase Ale Animalelor. Editura Brumar, Timisoara.

Morrison, R., Lee, J., 2003. Handling and sorting pigs in large groups housed in deep-litter systems. In: Second International Symposium on Swine Housing Held in Research Triangle Park, NC on October 12–15.

Nash, C., 1994. Update of the ACO Freedom farrowing system. NAC Pig Unit Newslett. 49, 4.

National Research Council, 2012. Nutrient Requirements of Swine: Eleventh Revised Edition. The National Academies Press, Washington, DC. https://doi.org/10.17226/13298.

Onan, G., Mircu, C., Patras, I., Hutu, I., 2016. Hoop structure for wean to finish pigs: management and gender influence in a summer trial. In: College of Agricultural Science and Veterinary Medicine, Iasi, RO: Veterinary Conference, fourth ed., vol. 59, pp. 388–394.

O'Connell, N.E., Beattie, V.E., Moss, B.W., 2002. Influence of social status on the welfare of sows in static and dynamic groups. Anim. Welf. 12, 239–249.

Payne, H., 2004. Deep-litter housing systems an Australian perspective. In: Hoop Barns & Bedded Systems for Livestock Production Conference, 14 September 2004, Ames, Iowa.

Payne, H.G., Nicholls, N., Davis, M., 2004. Group farrowing in a deep-litter eco-shelter. In: Alternative Swine Housing Systems: An International Scientific Symposium, September 15, 2004 Ensminger Room, Kildee Hall, Iowa State University Ames, Iowa.

Putten, G., Bure, R.G., 1997. Preparing gilts for group housing by increasing their social skills. Appl. Anim. Behav. Sci. 54, 173–183.

Renaudeau, D., Gourdine, J.L., Quiniou, N., Noblet, J., 2005. Feeding behaviour of lactating sows under hot conditions. Pig News Inf. 26 (1), 17N–22N.

Richard, T.L., Harmon, J., Honeyman, M., Creswell, J., 1998. Hoop structure bedding use, labor, bedding pack temperature, manure nutrient content, and nitrogen leaching potential, ASLR1499. In: 1997 ISU Swine Research Report, AS-638. Dept, of Animal Science, Iowa State University, Ames, IA, pp. 71–75.

Roese, G., Taylor, G., 2006. Basic Pig Husbandry - the Boar. Department of Primary Industries, State of New South Wales, NSW. Primefact 69.

Rossiter, L.T., Honeyman, M.S., 1997. Hoop Farrow-To-Finish Demonstration Iowa State University. Northwest Research Farm and Allee Demonstration Farm, ISRF97-29.31. http://www3.abe.iastate.edu/hoop_structures/swine/archives/other/systems.html.

Sällvik, K., Wejfeldt, B., 1993. Lower critical temperature for fattening pigs on deep straw bedding. In: Livestock Environment IV. Fourth International Symp., Univ. of Warwick, Coventry, England. ASAE, St. Joseph, MI, pp. 909−914.

Sas, E., 1996. In: Curs de zootehnie si etologie. Brumar, Timişoara.

Sas, E., Cernescu, H., Hutu, I., 2000a. Cercetări privind parametrii de reproductie la scroafele de rasă Hampshire. Lucr. şt. med. vet. Timişoara 31, 77−84.

Sas, E., Hutu, I., Florinda, F., 2000b. Ontogenia, senescenta şi performantele reproductive la scroafele de rasă Hampshire. Lucr. şt. med. vet. Timişoara 31, 85−89.

Selk, G.E., 2006. Management and Nutrition of the Bred Gilt and Sow, F-3653. Oklahoma Cooperative Extension Fact Sheets. http://www.osuextra.com.

Stalder, K.J., Johnson, C., Miller, D.P., Baas, T.J., Berry, N., Christian, A.E., Serenius, T.V., 2005. Pocket Guide for the Evaluation of Structural, Feet, Leg, and Reproductive Soundness in Replacement Gilts. National Pork Board. Pub. no. 04764, Des Moines, Iowa.

Svajgr, A.J., Hays, V.W., Cromwell, G.L., Dutt, R.H., 1974. Effect of lactation length duration on reproductive performance of sows. J. Anim. Sci. 38, 100−105.

Turner, S.P., Edwards, S.A., 2004. Housing immature domestic pigs in large social groups: implications for social organization in a hierarchical society. Appl. Anim. Behav. Sci. 87, 239−253.

Turner, S.P., Horgan, G.W., Edwards, S.A., 2001. Effect of social group size on aggressive behaviour between unacquainted domestic pigs. Appl. Anim. Behav. Sci. 74, 203−215.

Turner, S.P., Horgan, G.W., Edwards, S.A., 2003. Assessment of sub-grouping behaviour in pigs housed at different group sizes. Appl. Anim. Behav. Sci. 83, 291−302.

Wiegand, R.M., Gonyou, H.W., Curtis, S.E., 1994. Pen shape and size: effects on pig behavior and performance. Appl. Anim. Behav. Sci. 39, 49−61.

Young, M.G., Aherne, F.X., Main, R.G., Tokach, M.D., Goodband, R.D., Nelssen, J.L., Dritz, S.S., 2003. Comparison of three methods of feeding sows in gestation and the subsequent effects on lactation performance. Kansas Agricultural Experiment Station Research Reports 0 (10). https://doi.org/10.4148/2378-5977.6823.

Efficiency of swine production

The management system used for livestock production, whether conventional or alternative, and regardless of its specific advantages or shortcomings, will only be adopted if it is sustainable. The primary concern is economic sustainability, followed by environmental sustainability and finally social sustainability.

In Romania, current swine population and distribution indicates a robust recovery of the industry. After falling numbers during the years between 1990 and 2002, there are indications that the industry may return to the scope it enjoyed in the 1980s (14.8 million head in 1985). Particularly, encouraging is the large and increasing number of swine under private ownership. For example, in 2000, there were 5.8 million head or 6.3 livestock units/km^2, whereas by 2004, this had reached 6.5 million head or 7.1 livestock units/km^2 (Hutu, 2005).

6.1 Economic efficiency

Given their profitability and performance, alternative swine production systems (cold, deep bedded hoop structures vs. conventional confinement barns) have received praise from several types of farmers. One group of farmers that have successfully utilized this system are medium-sized farms with limited capital and plentiful labor that are capable of producing groups of 150—200 hogs during each production cycle.

Another group of farmers that can effectively utilize this type of system are those who are finishing hogs on a contract basis for large commercial swine companies.

From the standpoint of efficiency, raising swine in deep bedded cold barns has some advantages and disadvantages.

6.1.1 Advantages of deep bedded cold barn systems

The primary economic advantages of tarp-covered cold barns for swine production result from low capital investment and fixed costs and from some reduction in certain operational costs. Of all the fixed costs of swine production, those for construction and maintenance of the buildings stand out as the most significant.

Hoop barns are relatively cheaper and simpler to erect and can be easily constructed by the farmer (Onan et al., 2016; Hutu, 2005). For example, gestation units (without the specific gestation equipment) cost around 15,000 euro. Addition of feeding stalls raises the price to about 25,000 euro, but use of feeding stalls cuts total feed usage by 4.5% and allows for more effective control of sow body condition by allowing tailored feeding of each animal (Brewer et al., 1999).

Grow/finish units cost from 10,000 to 17,000 euro or about 50–85 euro per animal space, which is ¼ to ½ the cost of a conventional unit (Brumm et al., 1997; Kruger, 2006).

Some of the variable costs of operating a hoop housing system are also lower than conventional. These include reduced expenditure for electricity for lighting and ventilation, reduced fuel expense for heating, and lower property taxes and insurance fees. There is also the advantage of using home-grown grains and bedding materials for the system.

6.1.2 Disadvantages of deep bedded cold barn systems

Compared with conventional systems, hoop systems for swine do incur some increased variable costs as well as some losses in potential gross income.

Some of the operating costs are higher such as:

- higher daily feed intake during winter for swine to maintain body temperature;
- bedding costs are an added expense, especially if bedding is not produced on the farm and must be purchased;
- labor costs are higher as extra labor is required for distributing bedding and removing manure; there is also some increase in labor for sorting out sick animals and for adjustment of curtains and vents to maintain proper building temperatures and ventilation.

In addition, there may be some loss of income owing to slightly poorer carcass merit in hoop raised hogs. As most hogs are valued based on lean yield, the higher fat depths of hoop raised hogs in winter will decrease revenue as a result.

6.1.3 Cost comparison between hoop and conventional systems

Environmental conditions will significantly impact swine performance in cold hoop structures. Seasonal differences in growth rate, feed efficiency, and fat deposition are the largest factors (Morrical and Honeyman, 2005). In spite of this, cold barns do produce acceptable economic performance, even during cold seasons (Larson et al., 1999). Table 6.1 itemizes costs and returns for a hoop system versus a conventional system with all parameters averaged over an entire 1-year period (all seasons).

Compared with conventional systems, the hoop system's production costs are typically lower except for notable increases for bedding and for the increased feed consumption (5%–10% higher) observed in this system (Kruger, 2006). As indicated in Table 6.1, hoop-raised hogs provide a net return of $17.10 (USD) per hog over the entire year. This is only $3.46 less than the net return from conventional systems.

Table 6.1 Comparative economic analysis for grow and/or finish units.

Specification	Cold hoop barn	Conventional barn	Difference
Investment costs (in USD)			
Construction costs/animal	55.00	180.00	−125.00
Feeding and waste dump	36.00	36.00	±0.00
Total investment	91.00	216.00	−125.00
Cycles per year	2.91	2.68	+0.23
Total initial investment/cycle	31.29	80.48	−49.19
Tax, insurance, depreciation, %	16.5%	13.2%	+3.3%
Fixed costs/head marketed	*5.27*	*11.05*	*−5.78*
Operational costs			
Feeder pig purchase expense	38.00	38.00	±0.00
Mortality cost	0.77	1.52	−0.75
Insurance	1.31	1.32	−0.02
Fuel, repairs, maintenance	1.04	1.57	−0.54
Bedding expense	4.55	–	+4.55
Feed expense (at $0.132/kg)	37.95	37.23	+0.72
Veterinary expense	1.53	1.56	−0.03
Marketing expenses	1.53	1.56	−0.03
Labor expense	2.75	2.08	+0.67
Miscellaneous	0.77	0.70	+0.07
Total operational costs	*90.17*	*85.53*	*+4.64*
Total cost/head marketed	**95.43**	**96.58**	**(−1.14)**
Income from involuntary culls	0.11	0.00	−0.11
Net cost/head marketed	95.32	96.58	(−1.25)
Carcass value per animal	112.42	117.14	(−4.71)
Net income/head marketed	17.10	20.56	(−3.46)

Adapted after Onan, G., 2005. The application of hoop structures for growing and finishing swine, Lucr. şt. Med. Vet. Timişoara, 38, 342, University of Wisconsin, and The Economics of Finishing Pigs in Hoop Structures and Confinement Facilities: A Summer Comparison, 2002. Swine Research Report, 2001. Paper 18. http://lib.dr.iastate.edu/swinereports_2001/18, Iowa State University.

6.2 Sustainability

Economic viability and profitability are the primary focus for determining if a production system will work in any given situation. Today, however, it is also important that the production system be environmentally sustainable as well. A sustainable system is one whose production practices can endure for an unlimited time, i.e., have no long-term negative effect on the environment and use no significant quantity of nonrenewable resources. Honeyman (1991) suggests that a sustainable swine system includes a mix of production techniques that provide for a reasonable profit, have no negative environmental impact, and are acceptable to the society served. In Romania today, swine production sustainability may imply the following:

1. Low capital investment and financial risk. As indicated in Table 6.1, the low construction and fixed costs of a hoop structure will decrease financial risk compared with those of conventional systems (Gegner, 1997). Such structures are also multifunctional and may be readily converted to other uses either for livestock housing, equipment storage, or feed storage. These are, in short, low investment and flexible structures.

2. Optimizing value of farm products. Use of home-grown grains and bedding to grow swine can add value to these crops compared with what may be realized from cash sales.

3. Cycling of nutrients. Use of manure from hoop structures is an effective way to recycle nutrients back to the soil to enhance subsequent crop production. Manure is the most likely source for environmental damage if not properly managed (Baker et al., 1990), but when properly utilized and handled, it has very significant benefits such as adding fertility, increasing organic matter, reducing erosion, and enhancing microorganism growth within soils.

4. Animal welfare. Group housing on deep bedding allows animals a much wider avenue for expressing natural behaviors such as rooting. It also allows an animal more effective space and the opportunity to control its own microclimatic temperature. The use of hoop technology for gestation and farrowing, although intensive, reduces the stress on the sows compared with conventional systems (Hutu, 2005).

5. Reduced environmental impact on air, water, and soil quality. As rural population increases and a higher proportion of the residents are from nonfarm backgrounds, undesirable aspects of farm production become a larger issue. Of specific concern are negative effects on water quality and odor emissions from farms. Even though farms are often the economic drivers of the rural community, they nonetheless must accommodate the quality of life desires of other residents. Quality-of-life conflicts increase to an even higher level if the farms are large in scale. The perception is that larger farms cause more environmental damage (not necessarily the case). In such instances, manure is no longer fertilizer, but rather a disposal problem. Over applications of manure to soils can harm groundwater as can storage of large volumes in lagoons. Lagoons also often contribute to excessive ammonia and hydrogen sulfide gas releases, which result in a significant odor problem (Hogberg et al., 2005). Manure that is aerobically composted, however, as would be the case with hoop structures, is seen as environmentally responsible. In addition, reducing the total nitrogen load in the manure through carefully matching protein intake with requirements for various growth stages and through separate sex feeding is also seen as an environmentally friendly practice (Kruger, 2006).

In summary, environmental sustainability is a factor of growing importance in the swine industry and may soon reach the same level of significance as economic viability.

6.3 Social acceptance

Producing pigs in deep bedded systems is attractive for various reasons. The reductions in energy and other nonrenewable resource usage, the opportunity for establishment of niche markets as well as serving traditional markets, and the enhanced animal welfare are all desirable outcomes from the public's perspective (Fig. 6.1, Honeyman, 2005; Kruger, 2006). The market has a vested interest in this type of system because of its low production costs and low energy use. In addition, specific market segments, which have an interest in, and a marketing plan, which involves attainment of high levels of animal welfare, may use the advantages of the deep bedded system to meet the stated objectives and in their promotional materials.

Current survey data from Europe (Eurobarometer, 2005) indicates that only 5% of European pork consumers would ask for product where a "very good quality of life" has been available for the animals. Forty percent would request a "good life quality," and the remainder are unconcerned. The same survey indicates that at

FIGURE 6.1 Hoop structure at Experimental Units during 2016 Banat Agricultural Fair.

Promoting deep bedded swine production to Romanian farmers and to students at the Swine Experimental Unit.

Photo: Ioan Hutu, 2016, University of Agricultural Science and Veterinary Medicine, Timisoara, Romania.

shopping time, 43% of Europeans are also interested in the breed of swine (17% are always interested; 26% sometimes) as well as the animal's welfare. Another 20% are rarely interested in the breed or genetics of the hogs. A number of fast food restaurant chains are developing an interest in purchasing antibiotic free pork (Muirbead, 2003). In Europe, the market share for such meat is about 20%.

This is, in fact, a global trend as in the United States the organic food market increases by 20% per year and 40% of grocery buyers are at least interested in purchasing organically produced foods (Gentry et al., 2002). The state of Iowa, for example, has witnessed a significant increase in the number of consumers concerned about animal welfare friendly products. In light of these trends among consumers and the market, and the apparent willingness of at least some individuals to pay the extra price required for foods produced under more stringent rules, alternative animal production systems will likely have a bright future (Algers, 1997).

6.4 A sustainable swine production model

Given the above trends from Europe, Iowa, and Romania, one could envision a model system of alternative swine production with characteristics described in the remainder of this section (Honeyman, 1996; Hutu, 2005).

The base unit for the model would be a medium-sized family farm, which utilizes its own labor. These operations are not typically very intensive but would nonetheless have to meet all environmental standards for its peculiar location, including factors affected by topography, soil type, hydrology—both surface and underground, and air. Such farms will require expert technical assistance for swine production practices as well as environmental practices. These types of farmers are generally active stable community members. They are also likely to form production cooperatives and be proactive with responsible production practices and marketing.

For such a model to succeed well, it is imperative that these farmers have access to free current technological information regarding genetics, nutrition, management, and marketing of swine products. Young producers and beginners are a critical part of this model, and they, in particular, will oftentimes require extensive initial training in production practices as well as environmental and community issues. In addition, ongoing continuing education needs to be available to keep all producers abreast of current developments. Such education programs should encourage innovation for alternative management systems. Innovation is a prime mover of sustainable development.

Dissemination of innovative approaches and encouragement of further innovation will imbue a sense of sustainable responsibility in producers. In view of this, research priorities and demonstration projects should be focused on sustainable objectives and using inputs already available on the producers operations.

In this model, secondary products such as manure and dead animals can be viewed as sources of nutrients to be added to the soil or used for biogas production

FIGURE 6.2 North end of hoop structures.

Photo: Gary Onan, 2005, Wisconsin, River Falls, USA.

depending on the size and needs of the individual farm. Dead stock may be either composted or incinerated (Fig. 6.2).

Pork carcass standards in this model would use those of the current European quality controls and certifications either as pork (*SEUROP* system) or as traditional brands. For example, Mangalitza hogs, which are primarily lard producers, could nevertheless be promoted for their unique lean muscle sensory qualities (Hutu, 2003).

The 21st century demands that researchers and producers work out production systems that respect community values in addition to being productive, and environmentally and welfare friendly. It is imperative therefore that alternative system be explored fully and refined so that they can efficiently supply pork products while meeting the above constraints (Hutu, 2005; Hogberg et al., 2005).

In light of the current trends in alternative swine production, it is safe to predict that deep bedded hoop systems will become an important production system— perhaps one of many alternatives.

Considering Romania's tradition, available resources, and potential, there is a great deal of promise for the blossoming of such systems. This is a perspective shared not only by the authors of this book but by others as well. *Fiona Boal* said in 2006 *"With Romania's potential for significant grain production and cheaper labor, they have the potential to become the Iowa of Eastern Europe"* (Wacowich, 2006). Considering the social dimension of swine industry and future animal protein demand for human consumption, this statement may apply to many other developing countries equally well.

References

Algers, B., 1997. Cold housing and open housing - effects on swine health, management and production. In: 9th International Congress in Animal Hygiene, 17–21 August, 1997. Helsinki, Finland, vol. 2, pp. 525–527.

Baker, F.H., Busby, F.E., Raun, N.S., Yazman, J.A., 1990. The relationships and roles of animals in sustainable agriculture on sustainable farms. Prof. Anim. Sci. 6, 36–49.

Brewer, C.K., Honeyman, J.M., Penner, A., 1999. Cost of finishing pigs in hoop and confinement facilities Iowa State University Management/Economics. In: 1999 ISU Swine Research Report, ASL-R1686. Iowa State University, Ames.

Brumm, M.C., Harmon, J.D., Honeyman, M.S., Kliebenstein, J.B., 1997. Hoop Structures for Grow-Finish Swine. Iowa State Univ., Ames, IA. MidWest Plan Service, AED-41.

Eurobarometer, 2005. Report - Attitudes of Consumers towards the Welfare of Farmed Animals.

Gegner, L.E., 1997. Hooped Shelters for Finishing Hogs, Appropriate Technology Transfer for Rural Areas, Fayetteville, IN.

Gentry, J.G., McGlone, J.J., Blanton Jr., J.R., Miller, M.F., 2002. Alternative housing systems for pigs: influences on growth, composition, and pork quality. J. Anim. Sci. 80, 1781–1790.

Hogberg, M.G., Fales, S.L., Kirschenmann, F.L., Honeyman, M.S., Miranowski, J.A., Lasley, P., 2005. Interrelationships of animal agriculture, the environment, and rural communities. J. Anim. Sci. 83 (E.Suppl. l), E13–E17.

Honeyman, M.S., 1991. Sustainable swine production in the U.S. Corn Belt. Am. J. Altern. Agric. 6 (2), 63.

Honeyman, M.S., 1996. Sustainability issues of US swine production. J. Anim. Sci. 74, 1410, 1217.

Honeyman, M.S., 2005. Extensive bedded indoor and outdoor pig production systems in USA: current trends and effects on animal care and product quality. Livest. Prod. Sci. 94, 15–24.

Hutu, I., 2003. Research on Peculiarities of Reproduction of Sows in Intensive Management System. PhD thesis. Faculty of Veterinary Medicine, Timisoara.

Hutu, I., 2005. In: Mirton, Timişoara (Ed.), Reproducerea intensivă la scroafe.

Kruger, I., Taylor, G., Roese, G., Payne, H., 2006. Deep-litter Housing for Pigs, PRIMEFACT 68. NSW Department of Primary Industries, State of New South Wales. http://www.dpi.nsw.gov.au.

Larson, M.E., Honeyman, M.S., Penner, A.D., 1999. Performance of Finishing Pigs in Hoop Structures and Confinement during Summer and Winter. Iowa State Univ., Ames, pp. 102–104. Tech. Rep. ASL-R 1682.

Larson, B., Kliebenstein, J.B., Honeyman, M.S., Penner, A.D., 2002. The Economics of Finishing Pigs in Hoop Structures and Confinement Facilities: A Summer Comparison. Swine Research Report, 2001. Paper 18. http://lib.dr.iastate.edu/swinereports_2001/18.

Morrical, J.R., Honeyman, M.S., 2005. Serial Scanning of Pigs in Bedded Hoop and Confinement Buildings during Summer and Winter for Growth, Loin Muscle Area, and Backfat: A Progress Report. ISU Ext. Serv., Ames, IA. ASL-R2033. Animal Industry Report AS-651.

Muirhead, S., 2003. McDonald's establishes global policy on growth-promoting antibiotic use. Feedstuffs 75 (25), 1–4.

Onan, G., 2005. The application of hoop structures for growing and finishing swine. Lucr. şt. Med. Vet. Timişoara 38, 342.

Onan, G., Mircu, C., Patras, I., Hutu, I., 2016. Hoop structure for wean to finish pigs: management and gender influence in a summer trial. In: College of Agricultural Science and Veterinary Medicine, Iasi, RO: Veterinary Conference, vol. 59 (4), pp. 388–394.

Wacowich, J., 2006. Competition from Eastern Europe and beyond — new players in the global pork industry -summary of paper and presentation by Fiona Boal. Western Hog J. 27 (4), 40–42.

Index

143

Printed in the United States
By Bookmasters